THREE CELLS OF HONEYCOMB

GULES THREE CELLS OF HONEYCOMB SEALED
TWO AND ONE GOLD

The blazon of the arms of the armorial bearings
granted by The College of Arms, London
to Doctor Francis Godfrey Smith
of Nedlands, Western Australia

Each cell of honeycomb represents a continent
in which the author worked

THREE CELLS OF HONEYCOMB

by

FRANCIS G. SMITH
DSc (Abdn), BSc (Forestry), NDB, FRES

formerly
Junior Clerk, Lloyds Bank,
Gunner, Surrey Yeomanry,
Lieutenant, Royal Artillery
Beeswax Officer, Tanganyika
Senior Apiculturist, Western Australia
Director of National Parks, W.A.

Northern Bee Books

Three Cells of Honeycomb

Copyright © Francis G. Smith 1994

All rights reserved. No part of this publication may be reproduced, stored in a retrieval system, transmitted in any form or by any means electronic, mechanical, including photocopying, recording or otherwise without prior consent of the copyright holders.

ISBN 978-1-908904-41-6

First published in 1994 by
Dr Francis G. Smith
36 Vincent Street
Nedlands WA 6009
Australia

Published by Northern Bee Books 2013
Scout Bottom Farm
Mytholmroyd
Hebden Bridge HX7 5JS (UK)

Design and Artwork, D&P Design and Print

Printed by Lightning Source (UK)

CONTENTS

CHAPTER		PAGE
	Introduction	vii
	Acknowledgements	ix
1	1946: Into Forestry	1
2	Introduction to Beekeeping	11
3	Adulterated Beeswax	22
4	The Search for Proof	33
5	Daily Grind	40
6	Bee Botany	53
7	Completed Projects	67
8	Last Tour in Africa	81
9	1962: A New World	96
10	Legislation	107
11	Extension Work	113
12	Extracting Plant	123
13	Refinements in the Honey-House	134
14	Major Events	146
15	Eastern States Tour	161
16	Hives	169
17	Apiary Sites	184
18	Bee Breeding	190
19	Beeswax	199
20	Research	205
21	The Field for Improvement	218

Appendix:
	Method for research into bee botany	227
	Publications	242

INTRODUCTION

Beekeeping was neither my intended nor my chosen career: it all came about by chance.

I had made tentative moves towards keeping a hive of bees as a hobby in England in 1939 but the war put a stop to that. A practical start as a beekeeper was made in 1947 when I was living in the foothills of the Highlands in Aberdeenshire and needed to add to my income as a forestry student with a growing family. I read extensively, went on courses and eventually established a profitable little honey producing business. I also become involved in beekeeping research.

A clear intervention of Providence took place in a pub in England when the Director of Recruitment for the Colonial Office tapped me on the shoulder and asked me if I would like to do bees in Tanganyika. That invitation resulted in my becoming Beeswax Officer there for 13 years, earning myself the higher degree of Doctor of Science and being requested by leading publishers to write a couple of text books.

Then came independence for the African nation and the possibility of my having to seek employment elsewhere. I examined my prospects for a career in different parts of the world and targeted Australia in my New Year's resolution for 1962. The Department of Agriculture in Western Australia promptly offered me the job of Senior Apiculturist with a free hand to continue my research interests.

After nearly 12 years with the Department of Agriculture, the newly created position of Director became available in the organisation of the National Parks Board of Western Australia. I applied and I was appointed to the post, thus ending my professional involvement with beekeeping.

However, a few years after retiring, I was invited to go to Thailand on behalf of the Australian Government to advise the Thais on research in Apiculture.

Recently I received an urgent call to identify a colony of bees which had arrived at Fremantle attached to the underside of a container from Durban. I confirmed that the colony was of the common African honeybee, the 'killer bees' of the press, with which I had worked for so many years. Fortunately, the colony was destroyed before it sent out any swarms. From the amount of young brood and honey present, the African bees had been making the most of our coastal flora.

This account picks up the story of my life where I left it at the end of my earlier book, *One Gunner's War*. That was the tale of my experiences in the British Army from when I joined the Surrey Yeomanry as a Gunner in April 1939 until I was demobilised as a Commissioned Officer in 1946.

My wife Joan, whom I met in February 1939, was involved from the beginning. We declared our engagement on the 31st August of that year but the war and my absence overseas in the Middle East Forces delayed our marriage until March 1945.

Unfortunately she is allergic to bee stings. There was some evidence of this when we were in Scotland, but it manifested itself dramatically when we were in Tanganyika. So beekeeping then became a strictly professional occupation; there was no way it could be a hobby.

I have written this book because parts of it make an interesting story, particularly in the UK and Africa. I believe that what I was trying to do in Western Australia is still relevant today, and I hope that the articles describing the problems in WA can be of use to those who followed in my footsteps in beekeeping extension work. For this reason, I put no restriction on the use of material published in this book. I only ask that the source of anything which may be used be acknowledged in the usual manner.

Francis Smith
Nedlands 1994

ACKNOWLEDGEMENTS

There are so many to whom I owe a debt of gratitude for the contribution they have made towards my career that it would take another book to give adequate acknowledgement to them all. All I can do here is to mention those whose advice or actions had a decisive influence on the path of my life.

First and foremost is my wife Joan, who, from the day we met over 55 years ago has been my mate, my best friend, my supporter, my adviser and so very wise councilor.

Next was Professor H.M. Steven of Aberdeen, who picked me out of 400 other applicants for his 20 strong forestry course and later drew my attention to Professor Erdtman's work in Sweden when I was thinking about working for a higher degree. Then there was Alex Deans of Craibstone who enthusiasm and wide interests gave me much inspiration. Dr Eva Crane, who became Director of the Bee Research Association when it was founded in 1949, provided much help and personal encouragement to me for the whole of my bee career. I have to thank Murray Lunan who, as Provincial Agricultural Officer at Tabora, guided me through the intricacies of government department procedures and financial controls, and steered me on the path of extension work among African producers. Then there was John Groome, Provincial Forest Officer, who initiated the idea of the Beekeeping Division moving from Agriculture into Forests, and Dr W.J. Eggeling, followed by Robert Sangster as Chief Conservator, who gave me much active support.

In Western Australia, Dr Tom Dunne, Director of Agriculture and his Deputy, Leo Shier, greatly encouraged me during my early years in a rather difficult environment. Claud Toop, Chief Veterinary Surgeon and his successor, Dr Bill Gardiner, provided tremendous support as my immediate superiors. Bob Royce, Curator of the Western Australian Herbarium, assisted me by naming my collections of plants during the whole of my twelve years in the Department of Agriculture and Eric Lawson, Principal Adviser of the Information Branch, guided me in the production of my quarterly journal, *Apiculture*.

I am also grateful to Noel Fitzpatrick who, as a later Director of Agriculture, suggested that I take on vegetation mapping and gave me a leg up the salary scale. Ken Ashbolt, Chief Cartographer in the Department of Lands and Surveys taught me much about map-making and guided me in the compilation of material which could be converted into printed maps by his cartographers.

The above is a very limited list: everybody we meet, whoever they are, influences our lives to some degree. We interact with them all and learn from each other.

CHAPTER 1
1946: INTO FORESTRY

In May 1946 I was demobbed. For the past seven years I had served in the Royal Artillery, four and a half of them in the Middle East Forces. My service had concluded with a year with a regiment in the Hartz Mountains in Germany.

The abrupt transition from the army to civilian life was disconcerting. My wife, Joan, had made herself a little nest in the basement of a slightly bomb-damaged house in St Leonards-on-Sea near Hastings in Sussex. As a British high grade cypher operator she had been in General Eisenhower's SHAEF (Supreme Headquarters Allied Expeditionary Force) and then in General Montgomery's 21 Army Group Headquarters in Germany. She left the army a few months before me to have our first baby. Finding accommodation in Britain immediately after the war was not easy and her obtaining this flat had been nothing less than providential. But she was on her own ground; her grandparents had lived most of their lives in the area and as a child she had spent her holidays with them. Her 92 year old grandfather, Albert Grant, still sprightly, had the pleasure of seeing his first great-grandson.

While still in the army and having seen something of the work and life of officers in the German Forest Service, I applied to be admitted as a student at each of the four universities in the United Kingdom which offered degree courses in Forestry. Before leaving Germany, I spent a short time at Göttingen University studying the first year subjects. I knew that I would have to obtain some practical experience of forestry work before going to university, so as soon as I was back in England, I applied to the Forestry Commission for a job as a Forest Worker.

It happened quite by chance that the Forestry Commission wanted me to attend for interview on the same day that Joan and I had arranged to meet some old friends and go with them to a theatre in London. The Forestry Commission interview was to be at their Divisional Office in Woking, Surrey, and they wanted to see my wife also, no doubt to make sure that she would fit happily into an isolated rural community. We consulted the railway timetables and found that we could go to both the interview and our meeting in London on the same round trip, but there would be no possibility of our coming home to change between the two events.

So we duly reported at the Forestry Commission in Woking dressed for our trip to London, me in my army officer's Service Dress Uniform with Sam Browne belt and medal ribbons and Joan in a grey tailor-made suit with a white chiffon blouse which she had had made in Paris. The interviewing Board looked quite startled at our entrance but I got the job. The show we went to see in London was 'Arsenic and Old Lace'.

My work in the forest near Battle consisted of weeding young plantations of conifers. The tools for the job were a sickle, a sharpening stone and a stick. One gathered the 'weeds' together with the stick in one hand and cut them down with the sickle in the other. The weeds consisted of anything that ever grew in Sussex; grasses, herbs, shrubs and young trees. In some parts there was heather; that was hard to cut with a sickle – it was tough and springy. We each carried sharpening stones and made sure that the sickle was kept sharp: I appreciated the substantial structure and weight of the Government issue tools. In other parts of the forest we encountered the plants that grew in the ancient Forest of the Weald, including oak and hornbeam. These required different treatment and the tools for the job were a sickle in the left hand to gather the plants together and cut the softer stems, and a bill-hook in the right hand to cut the tough hardwoods. Hornbeam is so tough that the wood was used to make cog wheels in the mills.

I was not accustomed to such hard work and it took some weeks for me to be able to keep up with the rest of the team. In charge of operations was a Forester who visited us once a day to see how the work was going. He attended to all the paper work and paid us each week. In charge of the team itself was a tough and wiry countryman, the Ganger, with a Leading Hand to help him. The Forest Workers were a mixed bunch of countrymen and townies. Some were doing the work because they liked it, others because it was the only job they could get. But we all worked together as a team moving in line abreast through the plantation, each man to the space between two rows of small trees.

It was on this job that I learned that the most refreshing drink was hot tea, black, without sugar or milk. Joan would make up the most delicious sandwiches for 'breakfast' and 'dinner' breaks. To get to work I had to travel by train from St Leonards to Battle and them on my bicycle a few miles to the forest plantation. Between Hastings and St Leonards the railway went through a tunnel under the hills. We were living on one of the hills above the tunnel and could hear the trains when they went in at the Hastings end. As soon as we heard my train enter the tunnel I jumped on my bicycle and coasted down the hill to the station, arriving on the platform with my bike just as the train arrived; I put my bike in the guard's van for the journey to Battle. The only people on the train were regulars on their way to work and the driver and guard were the same each day. So when one day for some reason I was delayed and did not reach the station at the usual time, they held the train for a few minutes until I appeared. We started work in the forest at 7.30am.

The weather was good most of the time I was working in that Sussex forest though as summer approached it did get rather too hot for my liking. Light rain we ignored; we just kept on working. It was only if it had set in for a really heavy downpour by the time we arrived at the tool shed in the morning that we waited in the shelter for the rain to ease off before starting our walk to the work site. We got pretty wet as we pushed through the scrub but the hard work kept us warm and we did not suffer any harm from it, in fact I got in good trim for the next stage in my career.

The universities at both Oxford and Aberdeen invited me to attend for interviews. At Oxford I met Professor Champion in a building named RES RUSTICANA. The Professor was encouraging and Balliol College would accept me, but I was not too happy about Oxford. To start with there was an acute housing shortage and prospects for finding something nearby which would suit Joan and me and our baby Christopher seemed pretty remote. Then there was the climate. I had in the past found the humidity of the Thames valley trying, debilitating in summer and bone-searchingly cold in winter. Then there was the very serious matter of physics and chemistry. I would be expected to pass examinations in both these subjects before I started the forestry course. Physics I could possibly brush up but chemistry – I had not done it at school; the subject was banned after the science master blew up the school laboratory, and the little I had been able to do at Göttingen would not get me through.

Aberdeen was different. The sight of the granite city glistening in the early morning sun on my arrival was inspiring. I felt Aberdeen was welcoming. The restaurant in the railway station offered a beautiful breakfast of proper Scottish porridge, kippers and butteries, the Scottish version of croissants. I reported to the local equivalent of the Town Major and he examined my papers and gave me a certificate which identified me as an army officer on release leave and gave me entry to facilities reserved for military personnel. I was recommended to go and stay at Dormy House which was, to my great delight, occupied by others who were in Aberdeen for interviews for places on the forestry course. And the woman who ran the place, which was a sort of officer's transit camp and mess, introduced me to Scottish pancakes, which she seemed to produce endlessly on a hot plate, which she called a girdle.

My interview with Professor H. M. Steven in Marischal College was most cordial and I felt that I had made a good impression, even though, when he asked me what connections I had with the North East of Scotland, the only thing I could say was that my wife was a Grant (on her mother's side). A slight mile crossed his face. I was yet to appreciate fully the status of the name of Grant in the North East of Scotland.

I had also to be interviewed by the Forestry Commission's Conservator in Aberdeen to be approved as a person whom the Forestry Commission would employ when I got my degree. I think that was merely a formality to enable me to apply to the Scottish Education Department of a grant under the Further Education and Training Scheme for Exservicemen.

Both Oxford and Aberdeen offered me a place on their forestry course. I politely declined the Oxford offer and, with Dormy House as base, began to search for accommodation, combing the daily papers for anything possible. I could find nothing in Aberdeen itself so started searching further out, going in seemingly increasing circles from the city. Having no transport of my own, I was dependent upon public transport. I saw an advert for a likely place out at Tough and caught the first bus out. On arrival I was dismayed to see a fair number of cars and motor bikes already there. At one village, the postmistress suggested I enquire at a large house nearby, which she said had plenty of room, being occupied only by one elderly woman. I called at the house and the woman was most discouraging. As I walked back towards the centre of the village I was accosted by a policeman on a bicycle who wanted to see my identity papers. Apparently the woman had regarded me in my army uniform as a suspicious character and rung up the village bobby. Eventually, after many disappointments, I was directed to the house of Mrs Anderson in Torphins, a village on the Deeside railway, 22 miles from Aberdeen. Mrs Anderson, a plump cheerful woman, offered us a large room on the ground floor which we could use as our living room and another room upstairs for a bedroom. We also had the use of a shed in the quite extensive back garden in which we could store our small amount of furniture.

That settled, I was able to relax a bit and get myself organised for the move from St Leonards to Torphins and the prospect of three years' hard study. When our belongings were safely loaded into a furniture van, we went up to London to catch the Aberdonian night express and traveled first class in a comfortable sleeper compartment. On arrival at Aberdeen, we made for the station restaurant and had the anticipated excellent breakfast before travelling onwards on the Deeside railway to Torphins.

For the next three years that railway played an important part in our lives. It was a single line track between Aberdeen and Ballater, the terminus in the Highlands. There were passing places at the various stations where the driver handed over a large rigid brass loop to which was attached a bag which contained his authority to use the section of line, and obtained in exchange another bag for the next section. There were four trains each way each day. During term time I set off for Aberdeen on the first train in the morning and returned 12 hours later on the last train in the evening. In the winter I never saw Torphins in daylight except at weekends.

Apart from the spring gales, the summer had been fairly fine in the south of England. But in the north of Scotland the weather was unusually wet, too wet for harvesting much of the time. While I was searching the countryside for accommodation in August-September, I was aware of the dismal sight of stooks of oats which had been reaped and stood up to dry in the fields and were now rotting in the rain. That wet summer was a prelude to the hard winter which was to follow.

Our stay with Mrs Anderson was a valuable introduction to the way of life in rural Scotland. Mrs Anderson herself had lived in Torphins for 16 years, having come from the neighbouring village of Lumphanan, and she maintained that she was still regarded as an incomer. Food was still rationed but she taught Joan how to make the most of what was locally available and introduced her to a whole range of traditional Scottish dishes. The result was that we were able to live in some comfort on my student's allowance, at least for the first year; later the increase in the cost of living, and extra mouths to feed and bodies to clothe made it necessary to add to our income.

We had not been at Mrs Anderson's for many weeks when she told us that the County Council was going to erect some prefabricated houses just round the corner and she suggested that I should go and see the local representative on the Council, Inspector Laing, a retired policeman. He was most helpful and gave me some forms to fill in. There were various questions to be answered including number of children and years of service in the armed forces.

Very quickly ten Arcon prefabs were set up in a charming little cul-de-sac at the edge of the forest. To my great surprise and delight, Joan and I were offered number 5. It appeared that with our combined military service and our baby we had enough points to come fifth on the list. The council rents were well within what we could afford but we learnt that we would have to pay Feu Duty on the land to the Laird, Sir Thomas Innes of Learney. Actually that was nothing but a peppercorn rent.

The prefab was beautifully designed. In the centre of the house there was a solid fuel stove which heated hot water in a tank behind the stove and hot air from round the flue pipe was ducted to the two bedrooms. The plumbing was all, or almost all, in one unit wall with the kitchen on one side and the bathroom on the other. The exception was the pipe to the water cistern in the loo. That pipe ran from the main plumbing wall unit along the inside of the external wall under the bathroom window into the loo. As that wall was steel, the pipe was the weak point in the plumbing in winter time; to stop it from freezing we had to keep an oil lamp burning under it, a little oil lamp with a pink body and tiny chimney which Joan had taken to France with her during the war. In severe weather we used a Tilley lamp.

The kitchen was equipped with an electric cooker, an electric boiler for washing clothes and there was an ironing table which folded up against the wall when not in use. The pantry cupboard was rather too small for the country because we bought basics such as oatmeal and potatoes by the sack and there was no room for them. One other snag was that the house was designed to be run on electricity except for the solid fuel stove for heating. The Grampian hydro-electric supply had not yet arrived. But we had our Tilley lamps, a Primus and a Blueflam cooker as well as a Tilley radiator and a Tilley iron, all run on kerosene, so we were well equipped to cope until the Grampian eventually arrived some weeks later. Even then we kept the kerosene equipment on standby because storms and technical hitches made the electricity supply uncertain.

Joan's father had given us an oak bedroom suit. As the prefab had built-in wardrobes we did not need the wardrobe. I converted the oak panels at the ends of the wardrobe into solid bed ends, replacing the original vertical slats. What was over I made into a log box with decorative dovetailed corners for beside the fire, matching the dovetails on the oat hunting-chest I had brought from Germany. We furnished the rest of the house with utility furniture, a standard pattern of furniture which had been introduced as an economy measure during the war. As the same pattern was made by all furniture manufacturers, you could be lucky and get it with the highest standards of workmanship; the secret was to be able to identify the manufacturing marks of the best furniture makers. For floor covering in our bedroom we used a closely woven felt in a moss green, laid over a thick underfelt; in the living room and nursery we had carpets. With a few pictures on the walls, the prefab was a really cosy little home.

Fuel for the stove, such as anthracite and coal-based products, was rather expensive, especially when one needed to keep the fire going for months on end. However, birch logs were cheap and readily available; the trees grew like weeds.

I built a saw horse, and with a bushman's saw and an axe bought from Hector Kidd at the village store, we soon had a good supply of sawn and split logs stacked along the wall of the house ready for use.

I set about the garden, clearing the bracken and levering the boulders out of the black peaty soil with a wonderful tool called a tramp-pick. This was a strong crowbar with one end curved like half a pick axe, the other end with a wooden cross-bar handle, and there was a stirrup low on one side for pushing the point into the ground with one's foot. At the back and front of the house I planted lawn and the area at the side I turned into a kitchen garden. We had put down our roots in Torphins.

It was a short walk to the station and to the shops and Joan took Christopher out in a pram. But when the snow came and lay two or three feet deep on the ground the pram was useless. So I made a sledge like those I had seen in Germany, and fitted into it Christopher's little low wooden chair into which he could be securely strapped. For towing the sledge I made Joan a wooden pole with a leather strap to go round her wrist and a hook at the other end to secure it to a ring in the sledge. Out of one of my army blankets Joan made an all-in-one suit with a hood to keep Christopher warm. The sledge was highly successful and other people in the village copied the idea; the snow remained from November to April. There was only one mishap; Joan took a corner rather too fast and Christopher tipped over head first into a snow drift. He expressed his disapproval quite strongly.

Meantime my studies were progressing at the university. I revelled in the subjects of Botany and Zoology; Professors Matthews and Wynn-Edwards were superb lecturers and their assistants who took us for the laboratory work were excellent. Physics, Natural Philosophy it was called at Aberdeen, was considered a difficult subject but the recently appointed Professor R.V. Jones, formerly one of Churchill's war-time boffins and bender of German radio direction waves, had a unique approach to the subject which appealed to and was understood by his predominantly exservice students. He entered the lecture theatre where the students were assembled in their seats, contemplating the display of equipment on the demonstration bench. From beneath his black academic gown he produced an automatic pistol, took aim at something on the bench, and fired. A number of actions and reactions then took place among the equipment and the Professor proceeded to explain the processes and forces involved. Prof Jones also had a unique approach to examination. It consisted of numerous short questions designed to find out how much you know, not what you didn't know. I had no problems with Nat. Phil.

But the Chemistry! The head of the department was away, deeply involved with the development of the plastics industry. The lectures to the first year students, instead of being given by the Professor as was the Aberdeen custom, were delivered by a bored lecturer who merely read passages from the standard text book. The practical work in the laboratory was good and I was happy with that, but in the theory, I was lost. Not having done the subject at school was a severe disadvantage and I was not the only one floundering. Chemistry was the only subject I had difficulty with and I had to resit the end of year examination. I passed: I think the examiners were kind to those of us who were going on to other subjects.

CHAPTER 2
INTRODUCTION TO BEEKEEPING

Travelling from Torphins to Aberdeen and back each day on the train enabled me to do a couple of hours of study and revision in peace and quiet on the journey. I also got to know the other regular travellers. One of these was Alastair McCrae who kept a few hives of bees in his garden in Torphins. As a boy I had seen my great uncle Hugh Watkin producing heather honey on Dartmoor and in 1939 I had become involved in beekeeping to the extent of buying a hive and waiting for a chance to stock it with bees when the war intervened.

Alastair had large Glen hives, designed by Dr John Anderson of the North of Scotland College of Agriculture specifically for the climatic conditions in that part of the world. He introduced me to Gordon Hunter, a former Master Mariner turned commercial beekeeper, who had a small farm at the foot of the Hill of Fare, near Torphins. Alastair and Gordon agreed that a good time to buy bees would be next spring after the dreadful summer and hard winter we were experiencing; only healthy bees of good honey-producing strains would survive.

During the Christmas vacation I did some reading on bees; the Moir Library of the Scottish Beekeepers' Association was an excellent source of books. One book in particular impressed me; it was a slim volume called *The Art of Beekeeping* by William Hamilton, Lecturer in Beekeeping at the University of Leeds and formerly at the West of Scotland College of Agriculture. I came to realise that to practice commercial beekeeping with a large number of hives, and to be able to move the hives around the countryside to collect crops of honey at different times of the year and from different plants, only the simplest and lightest hives would be suitable, like those used in America, Australia and New Zealand.

These hives were based on the Langstroth design, which was not standard in the United Kingdom. I learned that a Scottish beekeeper by the name of W.W. Smith of Innerleithen had overcome the difficulty by designing a hive having all the features of the Langstroth hive but taking 12 British Standard frames with the top bars shortened. But first I needed to get some bees.

When I heard that a well known beekeeper further up the Dee Valley was having a roup (auction) to sell off most of his hives because of his advancing years, I went along and bought my first colony of bees. They were in a WBC hive, an English design, similar to but smaller than the Glen Hive, and a friend transported it for me to Torphins, where it was sited in pride of place in the centre of the lawn in front of our prefab.

We had acquired a cat, a grey cat with black stripes which we described as a granite cat and named it Binty – it was female. The cat was curious about the hive and the bees flying in and out and she went rather close to the entrance. Suddenly she took off vertically about three feet into the air and when she came down she fled. She had received her first bee sting. She did not take liberties with the bees again.

Joan and I familiarised ourselves with the bees in this hive. The queen even had a name, Agatha, Aggie for short. Christopher was now a toddler and he often squatted down near the entrance watching the bees fly in and out; if he was in the way, the bees queued up behind him, waiting for a chance to enter the hive with their loads of nectar. For some reason or other, he did not get stung. They were remarkably gentle bees.

As the colony increased in strength, we obtained some Smith hives from a manufacturer in England, and made increase. We now needed some place where we could keep the bees without their being a nuisance to the neighbours and where they would be in easy reach of the principal sources of nectar in the district. From study of the One-Inch Ordnance Survey Map of the area, I selected a shallow valley in which a stream ran down from the Hill of Fare and where the bees would be sheltered from the cold winds in winter time, get the benefit of the sun when it came out and be within reach of clover pasture, bell heather and ling. On my bicycle, I inspected the chosen spot and it was just as I had imagined. The little valley had a good downhill slope so it would not become a frost hollow, trapping cold and damp air; such places, although perhaps providing shelter from the cold winds, nevertheless are very harmful to the bees. The only difficulty was that the part of the valley I regarded as most suitable, lay within the boundary of a farm.

I found the farmer, a large biblical figure with a flowing beard, sowing oats by hand in awkward corners of a field, taking the seed from a kidney shaped wicker basket slung across his body and casting it with great sweeps of his arm. When I asked him if I could put my hives down by the stream, he gave his permission with a cheerful roar, "You can put them in the burn if you want to".

That site became our main apiary and winter quarters for the bees and we made a point of presenting the farmer and his family with a portion of the honey crop by way of rent. The farm was called 'Fordie', so we named our beekeeping enterprise 'Fordie Apiaries'. We took the hives there, and brought the honey crop back on a hurley, a two-wheeled barrow, which I made of wood and a pair of bicycle wheels.

It was a good summer and those whose bees had survived the winter did well. At the North of Scotland College of Agriculture's station at Craibstone, beekeeping courses were conducted by A.S.C. Deans, a very progressive lecturer and experimenter in a variety of fields. I attended his courses during the summer vacation, including one on the diagnosis and treatment of bee diseases. It was run to train disease inspectors under the Scottish Beekeepers' Association bee disease insurance scheme. The students on the courses were a broad cross section of society and included an ex-army Major who persistently referred to bee larvae as maggots, much to Alex Dean's annoyance.

Following the success of our first year in beekeeping – we had made a modest profit on our outlay from the sale of honey – we decided to increase the number of hives. Our experience of that year made us more selective about the source of new equipment. We had found that some manufacturers were making hives from timbers which were quite unsuitable for the job; there was too much movement with changes of moisture content. And they cost the same as hives made from the best timber, Western Red Cedar. The cheapest material for hive bodies was our own locally grown Scots Pine, which was perfectly satisfactory when properly seasoned and preserved on the outside with creosote and then thoroughly aired. I made some in Mrs Anderson's shed which I continued to rent from her as a store and workshop.

In between looking after the bees, attending courses at Craibstone, helping Joan with running the house and family, and swotting up on my failed chemistry, I managed to get in some practical forestry work, as required for the course, on the estate of Colonel Nichol of Ballogie. I worked with a very experienced woodman thinning stands of Scots Pine. Mostly it was axe work, but the bigger trees we felled with a two-handled crosscut saw. After felling, we cut off the branches and dragged the trees out of the forest to the side of a road. We worked at a steady pace and I did not find the work too hard; my civilian activities appeared to be keeping me in good trim. I also learned to understand, but not to speak, the Buchan, the local dialect.

The new academic year, the second year of our course, started us off on purely forestry subjects and we were now separated from the other departments such as Agriculture and Botany with which we had shared the same first year subjects. There were exciting new things such as forest botany, ecology, geology, forest engineering and forest utilization as well as silviculture. To my surprise I found that many of these subjects had some relevance to beekeeping. In addition to the ecological and environmental considerations, which dealt with the impact of forestry on bird and animal life, we also learned about the provision of amenity belts and other facilities for the enjoyment of the forest by the public, as well as the protection of the forest from pests and diseases.

At that time foresters were learning about and studying ecological relationships – the climate, geology, soils, topography , drainage, the interaction of various plant and animal communities and of course, man. And this was long before the generation of Greenies was even thought about. It is therefore not surprising that many years later, those best qualified to manage national parks were those who had trained in the UK to be professional foresters.

It had been a good summer, but by the beginning of December the snow had returned. We had so much snow that, on the night Joan found it necessary to go to the village hospital for the arrival of our daughter, Rosemary, the bus from Aberdeen had become stuck in a snowdrift and every car from the village, including the doctor's, had gone out to rescue the passengers. As she could not walk through the deep snow, Freddie Reid the fishmonger came to the rescue with his van. As he put it to Dr Morrison the next day, "You dinna ken I was at the relief of Ladysmith?"

Drying nappies with the temperature below freezing during the winter in the north of Scotland is quite a problem if your house does not have a drying loft. Arcon prefabs do not have drying lofts. So one has to put up with steaming nappies on a clothes horse in front of the fire in the living room. That also is the site for giving the baby his or her bath.

During that winter we were making plans for the management of the increased number of hives of bees. It would be worth hiring the railway lorry to take the bees round from one potential honey crop to the next. There were four possible crops apart from the variety of plants that contributed to the spring build-up: raspberries which were grown extensively further south, white clover in the local pastures and, on the grouse moors, bell heather followed by ling.

By now I had acquired an Enfield 125cc two-stroke motor bike with which I explored areas which I thought might be suitable for out-apiaries. I made an arrangement with a raspberry grower near Blairgowie in Strathmore to site my apiary in his field as soon as the canes began to flower. It was to our mutual advantage. He would get an improved crop of raspberries from better pollination while I would get an increase in colony strength plus some additional homey.

Early in June, the bees having built up well, I shut them up at night, having previously fitted ventilated travelling screens, and at dawn Geordie, the railway truck driver, helped me load the hives on his flat-bed lorry at Fordie and off we went to the rasps. Towards the end of June the flowering was almost finished and the hives were full of honey. I took off the full honey supers, put on empty supers and travelling screens and then had to wait for the bees to stop flying. It was not until midnight that it was dark enough for the bees to stay in the hives so that I could shut them in.

I then lay down on top of the stack of supers full of honey to get some sleep before Geordie turned up in the early morning. By three it was light again and a couple of hours later Geordie arrived. We took the bees back to Fordie to get the clover crop from nearby farms which had a lot of pasture, and then, in the third week in July, we were off to the heather on a well-managed grouse moor in the Forest of Birse. In spite of its name, the Forest of Birse contained no trees, although it had a potential for a good forest of Scots Pine.

The proper management of a grouse moor was important both for bees and for the grouse, which need the fresh growth of young heather for food, as well as the cover provided by the older bushes. If left too long, the heather grew tall and open, was of no use for the grouse and was unproductive for the bees. So to prevent this from happening the gamekeepers burned the old heather in patches. The patches burnt were no bigger than what the keepers and their helpers could control.

After the fire, the bell heather regenerated first and provided a honey flow for a year or two, and then the ling took over. The whole moor presented a mosaic of red bell heather and ling of different ages, and it was a dependable source of honey. In the Forest of Birse the valley runs more or less east and west, and flowering starts first on the south-facing slope and later on the north-facing slope, thus extending the season and improving the chances of there being at least some fine weather in which the bees can get out and collect a crop from the ling.

All this was well worth while and by the time we brought the bees back to Fordie for the winter, we had obtained an excellent crop. We were then faced with the problem of extracting it. We had extracted the raspberry and clover honey easily with the hand-operated extractor set up in the kitchen; and as it was summer, the weather was warm and the honey flowed readily.

But heather honey presented a different problem: it set in the cells of the honeycomb like a jelly, and could only be persuaded to flow by stirring. I acquired a device made of wood with many needles embedded in it at appropriate spacing. This we plunged into the comb to stir the honey in the individual cells before the comb went into the centrifugal extractor. Stirring the honey in the cells of the comb added to the work to be done. Another complication was that by the time the heather honey was harvested, it was now autumn and the weather was turning cold, so the kitchen had to be heated while we extracted and strained the honey. Joan and I solved the labour problem by inviting some of my fellow students out for a few days and supplying them with food and plenty of beer.

After bottling the honey crop, we than had to sell it. It had been a good year for honey production and all beekeepers in the north east of Scotland had more than they needed for their own use. So the local market for honey was flooded. Being quick off the mark we had sold our crops of raspberry and clover honey without difficulty, but the prized and potentially much more valuable heather honey looked like being difficult to move.

Joan now took a hand in the marketing. She advertised in *The Times* in London, "Scottish Heather Honey from Royal Deeside". That did the trick. We were inundated with orders from the south of England for more honey than we ourselves had produced. I was now able to buy heather honey of the best quality from other beekeepers in the district and give them a much better price than they could get otherwise.

We had our problems, the worst of which were damage to consignments which went to addresses south of London, and one customer put in a big repeat order and then failed to pay. When later we visited London, Joan and I watched the handling of parcels at the London railway stations. The Southern Railway was the worst; the porters hurled the parcels from the hand barrows on the platform to the far end of the parcels van, television sets included. No wonder we had breakages even with the special packing we had had made.

One of the local beekeepers who produced beautiful heather honey was a Lithuanian by the name of Lapoulski, locally referred to as 'the Ruski'. Somehow or other he arrived in the UK as a lad before the Great War. During the war he served as a Don R (motor bicycle dispatch rider) until one night he met a cow on the road and got a horn through his chest. His subsequent experiences made him a dedicated Communist in spite of his Lithuanian Catholic upbringing. At some stage in his career the condition of his lungs made it necessary for him to find a retreat in a healthy climate with plenty of fresh air. He settled high on the Hill of Learney above Torphins on a piece of land granted him by the Laird in exchange for some of his time looking after the Laird's bees. On this land he built three beautifully constructed wooden huts, one for sleeping, one for living and the other for cooking, and there he lived with his bees and his goats.

It was a good half mile from the road to his huts and the postie resented trudging up the rough tract to deliver loads of communist literature.

Joan and I met him in the village and knowing something of his background from Alastair McCrae, invited him in for a hot cup of tea on a cold winter's day. We found him interesting, in spite of his political persuasion, and from time to time invited him down for a meal and a chat in the evening.

On one occasion, he had got on his communist hobbyhorse and the argument was getting heated. Joan had picked up the iron tool which was used for shaking the ashes out of the stove when Lapoulski held up his hand and said, "Pleess – we will now talk about bees."

Another beekeeper who lived in the village was Harry Stephenson. I remember him greeting us ruefully one day in the village with the comment, "Ye ken, the kilt is no' just the ideal garment for the beekeeping." We made appropriate noises of sympathy.

The forestry course proceeded in a most pleasant manner. An important part of the course was the preparation of a working plan for a forest. This year the forest on the estate of Lord Cowdray at Echt was given as the subject of the project. This involved much work studying the habitat of each stand of trees, measuring, enumerating, sampling and thoroughly getting to know the whole of the forest in its various stages of growth. Then there was the matter of potential markets and the availability of equipment and mills to convert the thinning and mature trees into marketable products. Finally there was the not inconsiderable business of writing up the working plan and all its prescriptions for management in a clear and logical manner, together with the accompanying maps and plans, all hand drawn and coloured and all lettering done by hand, to me the most difficult part.

We also studied some of the forests on the royal estate up at Ballater, including digging soil pits – those of us who had been in the army were in our element as they resembled simple slit trenches, four or five feet deep. On one rather cool damp day, during our lunch break, one of the students cooked kippers on a spade over a discreet small fire at the bottom of his soil pit. Our lecturer, Willy McNeill, a retired Conservator from Ceylon which he called, "The best of all good shows", came striding through the forest exclaiming –

"The jungle simply reeks of kipper."

The winter of 1948-49 was not as severe as the previous two; we did not have as much snow but we did have icy cold winds which seemed to blow out of the heart of Russia. It was on a cold, frosty evening early in January that our third child indicated that it was time he came into the world. It was such a nice evening that Joan decided that she would walk round to the village hospital. So, carrying her bag of gear needed for such occasions, I escorted her to the hospital, without the drama of the 'Relief of Ladysmith'. The birth went well but after Joan had been home with David only a short while, we realised that we had a problem. For some reason or other, Joan ran out of her own milk and the baby was not able to digest the supplementary feed and was losing weight. The situation looked serious. Dr Morrison prepared us for the worst, "Some babies just don't make it."

But we were not satisfied with this. From somewhere we obtained an analysis of human milk and obtained the ingredients from the local chemist. One item was not available, a form of sugar. For this we substituted honey, which we had in abundance. We made up a formula as close as possible to the analysis. The result was dramatic. From the first feed on our formula, David started increasing in weight, and he never looked back.

Towards the end of our forestry course, we learned that, owing to financial stringency, the Forestry Commission had decided that it could not afford to pay increments in our salary for war service. To me that meant a very substantial reduction in my anticipated income. We also learned that in the Colonial Forest Service one could not take one's wife and family out on first tour. For this reason I had not applied for the Colonial Service. So when I heard that there was a research job coming up in the beekeeping branch of the North of Scotland College of Agriculture, which was housed in Marischal College in Aberdeen, I decided to apply for it and told Professor Steven of my interest and the reasons for it.

By this time I had joined the Bee Research Association which had been formed that year in England to meet the needs of all who were involved in bee research. So, together with my quite extensive reading on beekeeping, I was becoming well informed on the subject.

The last period of field work in the spring of 1949 took us down to the New Forest in the south of England and across to France to see the beautifully managed forests of Normandy. While we were down in the New Forest, the Director of Recruitment for the Colonial Office and the Forest Advisor to the Colonial Office were having informal interviews with those students who had applied for the Colonial Forest Service. One day, while we were having lunch of bread and cheese and beer in a New Forest pub called 'The Silent Woman' – the signboard showed her with her head tucked underneath her arm – I was deep in conversation with some other students when I felt a touch on the shoulder. I looked round, to see Mr Perry, the Colonial Office Director of Recruitment.

"Would you like to do bees in Tanganyika?" he said. I replied,

"I have a wife and three children."

"That's all right" he said, "You can take them with you."

"In that case I should like to hear more about it."

As soon as Mr Perry had gone, I turned to my companions,

"Where is Tanganyika?" No one knew for sure though someone did say that he thought it was about half-way up on the right side of Africa.

Tanganyika had a problem and the Colonial Office had an urgent need to find a graduate with beekeeping experience.

CHAPTER 3
ADULTERATED BEESWAX

Back in Torphins, a letter arrived from Mr Perry offering me a job after graduation, "to make a thorough study of African bees and to determine how native beekeeping, wax and honey production can be improved and stimulated" and, "initially full attention should be given to the improvement of the quality of beeswax exported." The position would be that of Beeswax Officer in the Tanganyika Department of Agriculture.

Joan and I gave it a lot of thought. It would mean uprooting ourselves from Torphins, which we were likely to have to do anyway, selling up our furniture and bees and setting off with our children into the altogether unknown. On the positive side, Joan would be able to get some help with running the home and looking after the children which, in spite of the convenience of the prefab, she was finding a heavy load. Further, the job seemed to have been made for me. It was one in which I would have a free rein to work as I thought best within the constraints of the terms of reference, financial limitations and government standing orders. We decided that I should accept the offer.

In replying to Mr Perry I suggested that I did some preparatory research in the UK, including getting some experience on a large scale bee farm. As the summer was getting on and without waiting for a reply from Mr Perry, I went down to Oxfordshire to work with R.O.B. Manley and Tony Rouse. Tony and his wife put me up at their place and I spent a valuable two or three weeks getting my hand in on a really large honey producing operation. I was already familiar with their methods from Manley's two books, *Honey Production in the British Isles* and *Honey Farming* but felt the need to see his techniques and novel ideas in action. I visited the headquarters of the Bee Research Association, which was in its infancy and was accommodated at Hull in the home of the Director, Dr Eva Crane. The library facilities and abstracting services as well as Dr Crane's personal encouragement were to prove most valuable to me then and in the years to come. I also visited the library of the Natural History Museum in South Kensington, which had a number of early references to African bees and their pests.

On the 3rd August 1949, Mr Perry wrote to tell me that he had heard from Tanganyika and that they would like me to visit the wax importers to find out what they wanted and the grades they were prepared to purchase, and also to visit the Imperial Institute to ascertain the correct firms to contact to acquire the necessary skill in preparation and grading, and any other firms the Imperial Institute might suggest.

The next two months were spent visiting and corresponding with the beeswax importers, the major users of beeswax and the refiners, who were able to give me a very clear picture of the trade in Tanganyika beeswax and of the difficulties which they were experiencing. Mr C.W. Gunn of Hale & Sons, Produce Brokers, was particularly clear on the relevant facts about the Dar es Salaam beeswax trade. It was he who drew my attention to a soft sticky wax which was appearing in consignments of Dar es Salaam beeswax. It was said to come from wasp's nests!

I learned that in Tanganyika, beekeepers and honey hunters collected combs from bee hives and from nests in hollow trees, and melted them down into cakes of wax of about two kilograms each. They sold these cakes in village markets to traders who in turn sold them to merchants and shippers in Dar es Salaam. To check for the presence of dirt and stones the shippers cut the cakes in half with a panga or machete. The nature of the fracture and the feel and texture of the wax indicated whether the beeswax had been adulterated by the addition of an oil, fat or other wax. Once mixed in, these materials cannot be separated out from beeswax.

Dar es Salaam beeswax was regarded as being generally of high quality, with room for some improvement in the preparation. The main problem was this 'wasp wax' which was slipping through into export consignments, in spite of sampling at the docks and analysis of samples by the Government Chemist in Dar. The 'wasp wax' made the beeswax sticky and useless for commercial purposes, even if present in small quantities. If it was not detected in time, it would ruin a batch of several tons of beeswax.

Messrs Salamon & Seaber, the consultant analysts in London, who regularly checked samples taken from beeswax shipments in UK ports, showed me some adulterated wax from Tanganyika, but they had been unable to determine the nature of its origin. At Redhill, in the workshops of British Wax Refining Co., Mr Arthur Case Green showed me a consignment of Dar es Salaam beeswax containing a quantity of bad wax at the bottom of some of the bags.

The chemists recommended me to get a book by A.H Warth which had been published two years earlier, *The Chemistry and Technology of Waxes*. From the description of beeswax adulterated with the wax of stingless bees in Central and South America, it appeared to me that there was a possibility that this sticky wax from Dar es Salaam was the same thing. The name 'wasp wax' could have come about because someone had learnt that the wax came from nests that had horizontal combs like those in a wasps' nest. However, wasps do not make wax.

In October, before leaving Scotland, I wrote a report on my UK investigations for the Director of Agriculture, Dar es Salaam, outlining the recommendations made by the various organisations I had visited and mentioning the problem of adulteration with 'wasp wax' and my theory as to its origin.

*   *   *

We already had the air line tickets from London to Dar es Salaam when Joan went down with chicken pox, caught on the Underground railway when we were in London kitting out for our new life in Tanganyika. She suffered rather badly with it and we had to wait until each of the three children in turn had developed symptoms and passed out of the infectious stage. A roup was held at the croft at the end of our road to sell our hives of bees and beekeeping equipment – a most distressing business. I packed my carpentry tools and a good supply of nails and screws in my old school tuck-box to follow us by sea with the rest of the transportable household effects. But I was not going to be separated from my all important text books.

Towards the end of November we were free from infection and again ready to travel. The baby, David was eleven months old and would have to travel in a carrycot and Rosemary, because she was two weeks under two years old, was also not entitled to a seat, in spite of her being able to walk. So she too had to have a carrycot. It had to be an outsize in carrycots to fit her and our village saddler, Saddler Reid – to distinguish him from Fish Reid and Shoe Reid – made one for her. This was the ideal container for my text books; neatly packed in the bottom and covered with a blanket and Rosemary could sleep in it if required. I did however notice a startled look on the face of the Air Hostess when she lifted it up into the aircraft at Southampton.

And so the day came when, having made arrangements for our furniture to be sold by roup, we handed over the keys of the prefab and, with our assorted baggage of army kitbags and valises, we boarded the train and said farewell to Torphins. From London we were taken by road to Lyndhurst in the New Forest where the airways put us up for the night in an hotel. Commander and Mrs Mulcahy came to see us off; they lived nearby. They were the parents of Maurice, one of the forestry students who helped in the extraction of our honey crops and who had been godfather to both Rosemary and David.

Early the next morning we boarded the Short Solent flying boat on Southampton Water. It had Pullman type accommodation, one pair of seats facing another pair with a table between them. The plane was not completely full and we had the four seats to ourselves, so Rosemary had a seat after all and baby David was comfortable in his carrycot. Joan was a seasoned air traveller, having done some eight journeys in DC3s during her army service; this was my first flight. For us seated near the front of the plane, the bow wave was a beautiful sight as it curled up over the window during the take-off.

That afternoon we landed in Sicily for refuelling and afternoon tea. From winter in the north of Scotland we had been projected into spring-like weather in the Mediterranean where daffodils were blooming. Later that evening we reached Alexandria in the dark, to be herded by soldiers with fixed bayonets and entertained by a gully-gully man with his tricks and sleight of hand, mostly involving eggs and day-old chicks.

We flew through the night towards Khartoum and from the plane Joan saw her first desert sunrise and was most impressed. We came down on the Nile to have breakfast ashore in the dry morning breeze before going on, following the White Nile up into the high country of Uganda with its great masses of cumulus clouds and turbulent air. These flying boats were not pressurised, so they flew low and one had magnificent views of the countryside and of herds of wild animals. The disadvantage was that in turbulent air they dropped into air pockets, upsetting the stomachs of some and causing the children to squeal with delight. Joan felt a bit queasy. Eventually, with the mountains of Kenya around us, we came down on Lake Naivasha in the Rift Valley.

Collecting our baggage before going through customs, we found that one item was missing, a cardboard box which, together with a hot box, had been put aboard the flight by the tropical outfitters in London. According to the invoices, the key of the hat box was in the missing cardboard box. The Customs were most understanding and they let us through without examining anything.

An open-sided truck with seats across the back took us up the rift wall to Nairobi where we were put into a chalet at the Norfolk Hotel, an old fashioned but comfortable place. There we experienced for the first time the customary East African call to dimmer, the British army mess call played on a xylophone by an African walking round the corridors of the hotel.

Joan and I had had no opportunity to change out of our Scottish winter clothing. Nairobi in the evening was cool enough but the next day we would be in Dr es Salaam on the Tanganyika coast and we would be needing lighter clothing. When we had put the children to bed, Joan and I sat on the floor of our room, trying to pick the lock of the hat box with a hair pin. Eventually it clicked and the lid opened. In it were Joan's clothes, three evening dresses, two afternoon dresses, a linen suit, and half a dozen frocks but no hat and none of my gear. So Joan was alright, and I still had the remnants of my army tropical uniform and could manage until we could get to the shops in Dar es Salaam.

Eventually we learned that the cardboard box had been taken off the plane at Alexandria and sent back to England by sea. It was then sent out to me in Tanganyika again by sea. It arrived many months later after I had re-equipped myself, and Joan's wide brimmed hat, packed in it, was squashed flat, beyond redemption.

Dar es Salaam towards the end of November was hot and sticky with the rains expected at any time. We flew there from Nairobi in a DC3 so familiar to Joan. The Customs in Dar es Salaam were much more officious than those in Nairobi and insisted in examining our modest possessions. We were met by a representative of the Department of Agriculture and put into the New Africa Hotel for a few days while I was introduced to various Government Officials, signed the Governor's book at the entrance to Government House and was briefed on my job. I also purchased essential items of clothing which I lacked.

Apart from the heat, which was very humid and to us was a bit much, we found Dar es Salaam to be a pleasant place, clean and spacious, with a quiet air of prosperity running through the whole community, African, Asian and European. The most important government building, the Secretariat, had a thatched roof and had been built in German times at the beginning of the century. The

Department of Agriculture was in a series of military type huts, thatched and whitewashed and all was clean, well kept, cool, and clearly very efficient. It is people, not ostentatious buildings, that produce a well run and contented society or government department.

My duties in Dar es Salaam completed, we boarded the mail train for Tabora which lay 36 hours westwards towards the heart of Africa. The train took us across the rich cultivated coastal plain, up the first escarpment to the beautiful mountains of Morogoro, across the dry Central Province with sparse flat-topped Acacia trees, scrubland with giant Baobabs and herds of well fed cattle, then up another escarpment to the central African plateau covered with 'miombo' open forest or woodland, full of wildlife, tsetse flies and honeybees.

Tabora was to be my headquarters, for this was the centre of the main area of beeswax and honey production, and the seat of the Provincial administration of the Western Province of Tanganyika.

The Provincial Agricultural Officer Murray Lunan and his wife Norah met us at the station to take us off to the hotel in their Bedford safari truck. That was the moment for the rains to arrive, breaking a two-year drought. In Tabora when it rained, it rained. The water just fell out of the sky and made the world fresh with the never-to-be-forgotten smell of the earth, newly wetted after the dry season.

We stayed some time at the Railway Hotel getting ourselves organised. The hotel was very comfortable and well run. The building was a former hunting lodge built for the Governor of German East Africa before the Great War. The children liked the African staff and quickly learned a few words of Swahili.

I was informed that the housing committee had allocated us some old RAF huts at the aerodrome, five miles out of town. As my job was a new one, there was no house for me to move into. Nor was there an office; I had to share a table with the District Agricultural Officer. There I got down to the task of learning government procedures, administration and accounting and the volume which ruled our lives, *Government Orders*. I also made a start on learning Swahili. I learned that each of the many tribes in Tanganyika had its own language: most related to a common Bantu origin, but others were very different, Nilotic, Bushman, even Caucasian. Swahili was the 'lingua franca' which was taught in schools all over the country.

RAMC hut, Tabora

Soon I was allocated an old wooden Royal Army Medical Corps hut next to the hospital. The roof was covered with black roofing felt which I coated with Kings Compo, a form of whitewash manufactured in Australia; it made the interior of the hut much cooler. I now had a room for myself and another for my staff and some storage space. I had a small allocation of funds which enabled me to employ a Messenger, Msema Kweli, who was a very young man when he joined me and was to serve me faithfully and be my constant companion on safari all the time I was in Tanganyika. I also took on a clerk who, because there was no government typewriter available, used my personal small portable. My staff was further enhanced by two

Beekeeping Instructors who had been employed by W.V. Harris, the former Government Entomologist at Morogoro, recently retired. I was able to borrow a simple dissecting microscope and laboratory utensils from other Departments.

While I was getting acquainted with the beeswax industry through the local Indian traders and African beekeepers, I sent the two Instructors out to find samples of the sticky wax, but they had no success. Victor Harris had been unable to collect enough of the wax of other insects for laboratory analysis, so considerable doubt was thrown on my theory about its origin.

A few weeks after my arrival in Tabora, a trader told me that in Central Province there was a quantity of sticky wax which had been returned by a shipper in Dar es Salaam. I went by train to Manyoni and found the wax in a trader's store. There were some 200 kilograms of the stuff; some a mixture of beeswax and 'wasp wax' and some seemed entirely 'wasp wax'. I had no authority to seize it so I bought a two kilogram cake of the latter and returned to Tabora.

Back in my office I tried to cut the cake in two with a panga. One good blow with a panga would break a cake of beeswax in two easily. But this material stuck to the panga and when I did succeed in cutting it in two, the gleaming sticky cut surface indicated that this was indeed 'wasp wax'. And, to my great delight, I could see the bodies of small insects embedded in the wax. After melting and straining the wax and cleaning the residue with solvent, I identified the insects as stingless bees, family Meliponidae, probably of the genus *Trigona*.

I returned to Central Province and toured the trading centres, searching the stores of the beeswax traders for sticky wax. From seven different villages I collected samples. From these I obtained part of the bodies of stingless bees; sometimes only the mandibles and hind tibia were identifiable but this was enough for me to distinguish two separate species.

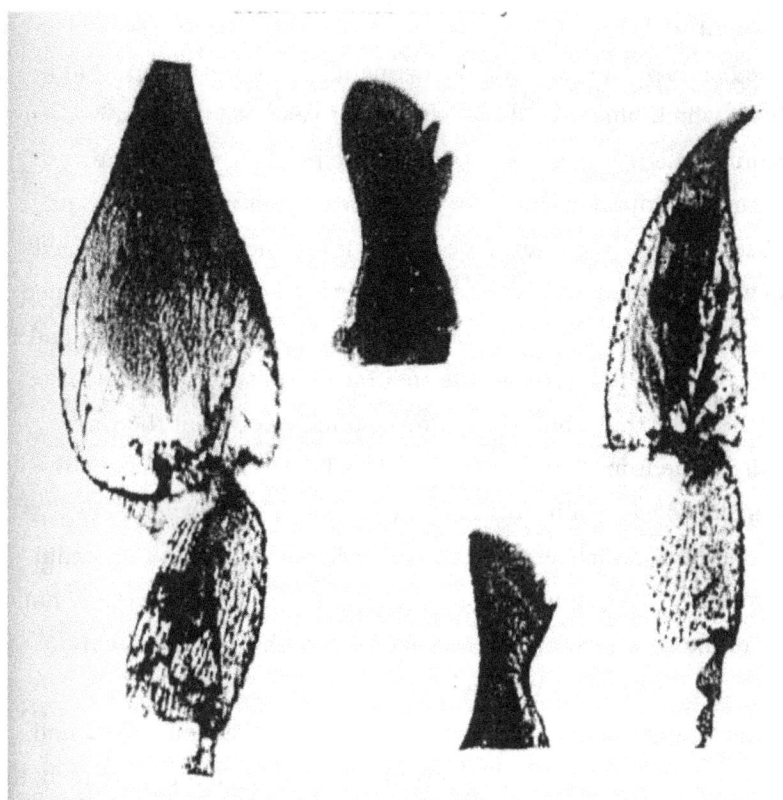

Hind tibia and mandibles of Trigona bees found in cakes of adulterated beeswax

My collection of wax samples went off to Dr W.D. Raymond, the Government Chemist. I also sent him some samples I had concocted of mixtures of pure beeswax and Trigona wax in carefully weighed proportions. I wanted to know how readily the presence of the sticky wax could be detected by chemical analysis.

Dr Raymond's results showed very different values for beeswax and for Trigona wax in each of six different chemical properties. In my specially prepared mixtures, the presence of Trigona wax could be detected down to as little as five per cent.

By September 1950 I had sufficient data to publish a preliminary report on Trigona wax, which appeared in the *East African Agricultural*

Journal, April 1951.

The heads of Government Departments could not accept that a problem which had baffled the experts for years had been solved by a newcomer in a few months. It was pointed out to me that the Government Entomologist, with all the resources at his command, had been unable to obtain more than a few ounces of the so-called wasp wax. Dr Raymond questioned the evidence on which I claimed that certain samples were Trigona wax and enquired whether I had actually obtained Trigona wax from a hive or nest.

It was clear that I had yet to prove conclusively that the wax actually came from the nests of stingless bees. The presence of the bodies of *Trigona* in a number of samples of the wax was merely circumstantial evidence. So my next task was to obtain the actual nests of those bees; it was not going to be easy. As Dr Raymond had told me, the vastly experienced Government Entomologist, with all the resources at his command, had not succeeded.

I was to appreciate later that the Government Entomologist had been badly let down by 'all the resources at his command', namely the two Beekeeping Instructors he had transferred to me. I found them to be utterly useless.

CHAPTER 4
THE SEARCH FOR PROOF

Beekeepers and honey hunters in both Western and Central Provinces told me that the nests of various stingless bees occur in hollow trees and also in the ground, some of them in termite nests. There were also some very small species, 'mpunzi', which were very common and which made nests in small hollows under bark and even in key holes in doors. These diminutive bees have the annoying habit of hovering round the eyes and settling on arms and legs in search of moisture, hence their common name, 'sweat bees'. They produce wax for the construction of their nests but the quantities are so small that they were clearly not the source of the wax which was causing problems in the trade. Shippers in Dar es Salaam were now telling me that one and a half tons of 'wasp wax' was passing from hand to hand among the Indian merchants in Tanganyika.

The species of *Trigona* whose nests I was seeking were about the size of a small house fly or a bush fly. The only sign of the presence of a nest was a small wax entrance, sometimes trumpet-shaped, protruding a centimetre or so from the trunk of a tree or the surface of the ground. It was so small that only one or two bees could pass in or out at one time. In the vast African bush it was like looking for a needle in a haystack.

By this time I had my own Land Rover and my old army camping gear had arrived, so I was able to travel about the country freely. I visited the villages where Trigona wax had appeared. I camped at the Government Rest Houses and told the local Native Authorities that I was looking for the nests and I offered a reward for each nest anyone could show me. While waiting for the news to get around, I myself searched in the bush, catching bees on flowers and looking for the entrances to nests.

I made it clear that it was most important that I be taken to the nests: I did not want the nests to be brought to me. I needed to see how the nests were sited and to record the details of their construction. It was essential that I be present while each nest was cut out of a tree or dug out of the ground. As information came in, I found that none of the nests was easily accessible. The tracks were rarely used or non-existent and much bush walking was involved. It appeared that all the easily accessible nests had already been destroyed for their wax and honey.

Gradually patience and persistence were rewarded and I was shown the nests of three distinct species of the larger *Trigona* bees. During two weeks in Central Province in June-July 1951 I collected 25 nests. One species made nests in hollow trees, another in cavities in the ground close to the surface and the third in termite nests, two or three metres underground. Once the little waxen trumpet of the entrance of this last was discovered, digging began, carefully following the cerumen lined tube of the entrance tunnel down through the soil until the black football-sized outer coating of the nest was reached. The soil, the hard fine material of the termite mound deep underground, was carefully removed from one side of the nest. This revealed also the galleries of the termite nest, for this species of *Trigona* always built alongside the brood galleries of the termites. Once the shape of the Trigona nest was determined, and its drainage tunnel identified, the outer case and its labyrinth of wax plates was cut away to reveal the brood combs and the honey and pollen storage pots.

There was no fighting between the termites and the stingless bees; if left alone they each built up their own kind of walls, the termites' of damp earth, the bees' of wax. I took photographs of the nests with a small old vest-pocket Agfa camera; exposure under these conditions was a matter of guesswork and I tended to over expose. The amount of wax extractable from each nest varied from 30 grams to one kilogram and these samples I sent to the Government Chemist of analysis.

Nest of Trigona denoiti in termite nest.
Outer cover removed to show honey pots.

Earlier in the year I had heard from H.J. de S. Disney, Entomologist at Ilonga in the Easter Province, that he could show me the nests of another species of *Trigona* which makes a nest in the open under a branch of a tree. I was not in a position to go myself at the time so I sent one of the Beekeeping Instructors I had inherited from the Government Entomologist to examine the nests and collect specimens of the bees. His trip was not successful because, according to his report, the bees had attacked him with great ferocity, biting him and getting into his ears and nose. He did not explain why he did not wear his government issue bee veil..

Although I had no evidence that the wax of this species had been collected or used to adulterate beeswax, it was desirable to see how its chemical and physical properties compared with the wax of the other three species.

As soon as I could, I went to Ilonga with Msema Kweli and the Beekeeping Instructor, all three of us properly attired in protective clothing.

Nest of Dactylurina staudingeri at Ilonga

Smoke had little or no effect on these thin-bodied bees; they attacked all the time. While they could not bite us, having those furious little creatures dancing on the outside of the bee veil an inch or two from nose and ears was rather disconcerting. In addition to the bees we had to contend with the green tree ants, called 'maji moto' (hot water), which made their nest in a cluster of leaves, pulled together and secured in some way at the end of a branch. There was also a climbing plant which had a large bean-like fruit which grew in pairs, shaped something like the horns of an African buffalo, hence the name 'buffalo bean'. These beans had hairs which easily detached and caused considerable irritation.

There was a story that a team of British army surveyors, working under these difficult conditions, received complaints about the condition of their field note books, which they sent periodically to the mapping branch in England. In reply, between the pages of the next set of field note books, they included hairs of the buffalo beans. Complaints ceased.

To escape the worst of these hazards, I took the nest into the shelter of the rest house to examine it, but still our bee veils were needed. My pencil, notebook, camera and our protective clothing became covered with small blobs of sticky wax, undoubtedly a most effective weapon against their major enemies, the aggressive tree-dwelling ants.

While the other three species of *Trigona* made horizontal or nearly horizontal single-sided brood combs, the new species built vertical double-sided combs for its brood. It stored its honey and pollen in wax pots like all the other species of Meliponidae.

After my examination, I put this nest into a cloth bag to prevent any bees escaping into the Land Rover and drove down to Dar es Salaam to present this, my final example of *Trigona* nest material, to the new Government Chemist, W.E. Calton; the critical and doubting Dr Raymond had retired. Opening the bag in the laboratory caused consternation. The African staff disappeared but Mr Calton stood his ground, fascinated, sticky bees embedded in his beard.

When the chemical and physical analysis of the wax from the nests of the four species of *Trigona* was completed, my early theory that 'wasp wax' was from stingless bees was proven beyond doubt.

Next I needed to be able to put names to the species before publication of the results. This proved more difficult than I expected. The various authorities in entomology I consulted even went as far as indicating that no one in the world could name the different species of African *Trigona* with certainly. However Herbert F. Swartz of the American Museum of Natural History, New York, offered to check my specimens with the collection in his museum, and he provided me with names for each of the four species.

The first, which makes nests in hollow trees, he named *Trigona togoensis*, Stadelmann, but this was later identified as *Trigona ferruginea*, Lepeletier 1836.

The second, which makes nests close under the surface of the ground, was *Trigona beccarii*, Gribodo 1879.

The third, which nests much deeper in the ground, usually in association with a termite nest, was *Trigona denoiti*, Vachal 1903.

The fourth, the thin-bodied species which makes its nest under the branch of a tree and has vertical double-sided brood comb, was named as *Trigona staudingeri*, Gribodo, later *Dactylurina staudingeri*, (Gribodo) Cockerell 1934.

I completed the writing up of the results of my investigations in November 1951. General descriptions of the four species and their nests were published in the *East African Agricultural Journal* in October 1952 and a complete description of the nests and chemical and physical properties of the waxes was published in June 1954 as 'Notes on the Biology and Waxes of four African Trigona Bees' in *Proceedings Series A of the Royal Entomological Society of London.*

There now remained the application of the results of my work in a practical form that would benefit the beeswax trade. The existing *Produce Export (Beeswax) Rules* were clearly ineffective. The analysis of the samples taken by Customs at the port of Dar es Salaam was not done until the consignment was already in the ship on its way to England. As all beeswax produced in Tanganyika was exported and many merchants were involved, the obvious point for the enforcement of standards was immediately prior to loading on the ships. I set about revising the Beeswax Rules, tightening the specifications to permit only pure clean beeswax to pass. Inspection of the consignments was to be carried out by Customs Officers as before, but not only were the bags to be opened, some of them were to be tipped out on the floor to ensure that there was no adulterated material hidden at the bottom of the bags as had occurred in the past.

I taught the Customs Officers to recognise dirty or adulterated wax by simple inspection. If they found doubtful wax they could send samples to the Government Chemist and detain the consignment until the results of the analysis were received. This relieved the Chemist of detailed analysis of every consignment. Analysis would be required only if the Customs had doubts about the wax. If the analysis showed that the Customs Officer was correct in his suspicion of adulteration, the whole consignment would be forfeited and the shipper fined three times the value of the consignment for making a false declaration to Customs.

It took the Tanganyika Government until 1957 to pass the revised Rules. By then the export trade was bedevilled not only with Trigona wax and ordinary dirt; adulteration with cheap and readily available paraffin wax was taking place on a large scale. The merchants who were profiting from the sale of adulterated wax complained bitterly about the proposed new standards and succeeded in delaying acceptance by Government for several years.

I had trained 100 Beekeeping Instructors, some employed by Government, others by Local Authorities, all working among the producers, teaching them how to prepare cakes of clean beeswax. And still unscrupulous merchants continued buying adulterated wax and were successful in exporting it under the old rules. This enabled it to be sold to brokers while on the high seas, vouched for by the Government Chemist's Certificate of Analysis. As I crudely put it, "If there is a market for shit, then shit will be produced."

I was the recipient of the complaints from the overseas refiners and manufacturers.

When at last *The Produce Export (Beeswax) Rules, 1957* came into effect on the 1st February 1957, it took the application of the penalties against only one merchant for the whole trade, shippers, merchants and up-country traders, to realise the game was up and to refuse to buy any more adulterated beeswax.

Seven years after my arrival, I had achieved the prime objective of my appointment – the improvement of the quality of beeswax exported and the elimination of 'wasp wax' from the trade.

CHAPTER 5
DAILY GRIND

Many other things happened during those first seven years at Tabora while I was trying to overcome the adulteration of beeswax, some seriously threatening the well-being of the family.

During the rains when we were living in the RAF huts at the aerodrome, the Public Works Department had a night watchman stationed there. His child played with our children until one day we discovered that both he and his child had leprosy. Joan, who was then sickening with her first attack of malaria, expressed her wrath to Dr Kevill, the Senior Medical Officer. The Housing Committee hastily found us a house in Tabora. It was a nice house built between the wars and we were very content with it.

But not many months after moving in and getting our things organized, Murray Lunan went on leave and we were asked to take over his house to keep it in the Agricultural Department. When Tabora was the German military capital it had been built as the Sergeants Mess; it was solid, cool, roomy and seemed ideal for our growing family. It also had a beautiful old muninga table of which the Lunans were very proud. Between the Lunans leaving and our moving in, the table disappeared. Joan checked all likely households but there was no trace of the table. But before the tour was over we were allocated, on the outskirts of the town, a fairly new house which had been occupied previously by a police officer. When we walked in, the first thing we saw was the Lunan's missing table. We looked after that for the rest of our years in Tabora.

Unlike most houses in Tabora this new house, having been occupied by a police officer, had a telephone which I arranged to be connected by a direct line to my newly built research station. This house was much more convenient to run and remained as the

Beeswax Officer's house for the rest of our stay in Tabora.

There was usually a break in the rains in January so I arranged a safari to Mpanda, going down on the goods train and returning by road through the important beekeeping centres of Uruwira and Inyonga. The branch line to Mpanda was fairly new and, unlike the main line, the railway sleepers were not of steel and bearing the stamp of Krupps 1906, but were of treated local timber. As the train crept cautiously along, jets of water shot out of the ends of each sleeper. We reached Mpanda safely but a later train went missing. An aeroplane sent out to look for it, found the locomotive lying on its side in the black clay of the flood plain of the Ugalla River. No one was hurt and when the plane flew overhead the train crew and handful of passengers were seen to be cooking a meal.

My visits to Uruwira and Inyonga were very successful in the contacts I made with the beekeepers and the traders. At Uruwira I was entertained with cups of rather too sweet tea by the principal Arab trader and by the time I reached Inyonga I was not feeling quite myself. A very kind Indian trader at Inyonga, aware that I was sick, made sure I had everything I needed for the return trip. That night heavy rain fell.

My enquiries revealed that a trader's truck had just returned from the Ugalla and the river was still passable. Feeling that I should get back to Tabora as soon as possible to get proper medical attention, I set off in the Land Rover accompanied by Msema Kweli and Hamedi. All went well for the 45 miles up to the Ugalla River; the sandy road was reasonably dry. There was water in the river and on its surface floated masses of water lilies. Msema and Hamedi waded across and found the water was shallow enough for the vehicle to ford, but just in case, they stood on either side of the track in the deepest part ready to push. The Land Rover went through and up the opposite bank without any hesitation.

Once clear of the riverine vegetation on the north bank, we could see that the grass on the flood plain was growing up through water. I drove forward cautiously, the track being clearly indicated by the different kind of grass growing on it. We expected the track to rise above the water level at any moment but on and on we went, still through water. We travelled 22 miles along that flooded track before we reached dry ground in the miombo forest. I was exhausted with the strain of the driving but I pushed on and completed the 130 miles from Inyonga to Tabora. Caked with mud we arrived home and I collapsed on a settee. Joan called the MO, Dr Caine, and he diagnosed glandular fever and ordered me to bed. I still have nightmares about driving on flooded bush tracks.

The occurrence of the wet and dry seasons was normally fairly predictable, so I was able to plan safaris to avoid the rains. But I got caught on another occasion. A carefully planned trip round the north and west of Western Province in May, when the dry season should have begun, coincided with the arrival of a cyclone in the south of Tanganyika. The wind was severe enough in Western Province to bring down many beehives just when the major honeyflow was taking place. It also brought rain. When I reached the Malagarasi River half way between Kibondo and Kasulu, I found it was impassable for motor vehicles. I could have boarded the ferry on the steep north bank, but on the south side the ferry stopped on the flood plain with its ramp in three or four feet of water. I had to return to Kibondo and send a telegram to the District Commissioner at Kasulu saying that I could not arrive as planned that day. I did not know it at the time but it took ten days for the telegram to travel the 90 miles from Kibondo to Kasulu. Naturally, the DC was a bit peeved at my failure to appear as arranged.

There were also problems with my return to Tabora by the route I had come; rivers had risen and were flowing over bridges. Reconnaissance showed that numerous traders' trucks were waiting on both sides of the first river. Not wishing to share the tsetse flies and mosquitoes, I returned ten or fifteen miles to a rest house to wait a couple of days for the traffic to start flowing. When the trucks had passed through, I set off for the river, but on arrival I found the water level over the bridge was still too deep to drive the little Land Rover. The actual flow of the river was quite gentle so we towed and pushed the vehicle over by hand.

The sodden state of the road, which was nothing more than a bush track at the best of times, caused so much drag on the wheels that the petrol consumption was greatly increased. Five miles from Kahama, I ran out of petrol. Hamedi and Msema took a jerrican into the town while I waited in the vehicle. Hours later they reappeared out of the darkness with the petrol, having wisely had a meal while they were in the town. That is the one and only time I have ever run out of fuel.

Back at Tabora, I learned that my office in the RAMC nut had been burgled and a quantity of beeswax taken. This wax had been stored in the bathroom in the end of the hut behind my desk. The wire in the bathroom window had been cut and the shutter opened so it looked as though the thief had come in that way. However, Msema told me that while we were held up by the floods, he had a dream about my chair, and he had been afraid that someone had stolen it. Actually that chair was a very simple one, made entirely of muninga, a beautiful African mahogany, and it had a carved seat. It was much admired as a piece of furniture and Unyamwesi craftsmanship. To get into the bathroom from my office, the thief would have had to move my chair. We had the unpleasant feeling that the burglary was an inside job.

The police made enquiries, but they were baffled. At the suggestion of Hamedi and Msema, I enlisted the help of the teacher at the Mosque. Joan tells the story of what happened then in *A Patch of Africa* under the title of 'The Alternative'. The matter was resolved in a rather wonderful African way, though I never got the beeswax back.

There were other dramas. Joan got stung by a bee, just one sting on the finger when, ignoring the warnings of Hamedi, she was investigating the presence of a swarm of bees in the old ice box on the back verandah. She developed an acute reaction and, recognising what was happening, she managed to get herself to the hospital in the

Land Rover, driving supported by Susanna, the ayah, the children's nurse maid. When I was told at the office about Joan having been stung, I went straight to the hospital with my own supply of adrenalin. On hearing that the hospital's adrenalin had had no effect I asked to see it; it was dark and cloudy. I asked that Joan be given another injection, this time with my fresh adrenalin. This was administered only after I had a fierce argument with the Medical Officer, Dr McGuiness, who maintained that there was nothing wrong with the hospital adrenalin;

"She is my patient" he said.

"She is my wife" I replied. I got my way. The effect of the fresh adrenalin was immediate.

It happed that that was the day of the Caledonian Ball for which Joan had been preparing for some time and for which we had become very competent in all the dances. Joan recovered in time to hear the music of the ball from her hospital bed. She asked me from where the music was coming. I told her and she wept. I did not go to the ball either.

Before we went on leave, Dr John Craddock, the new MO, gave Joan a letter of introduction to Dr A.W. Frankland, the expert on bee venom hypersensitivity at St Mary's, Paddington, so that a desensitising kit could be prepared for her.

We expected a baby to arrive at Christmas 1951 and the nurses even decorated the ward at the small European hospital for the event. But on the 25th February 1952, the current Medical Officer, Dr Hanley Laycock, decided that time had come to act, so Joan was admitted to hospital and later that day gave birth to our fourth child, Antony, weight nine pounds plus. I was with Joan until the moment of birth and five minutes later the MO took a photo of the baby in my arms.

The baby was still very small and had not begun to crawl when we were in the centre of a rabies outbreak. Our dog had been bitten by its father, who then disappeared. With no thought about rabies, Joan cleaned out the wound with neat Dettol. It was not until some time later that it came to light that the dog's father and the mother had both developed rabies, and that the male dog had bitten all the other dogs along the street. All were put into kennels for observation, including our dog, and eventually all the dogs, except one, developed rabies. The exception was ours, treated with Dettol.

The families who had dogs which had been bitten were considered contacts and had to undergo a series of injections, two a day, in the stomach area, for 14 days. The first week was not too bad, but the second week was most painful. The frightful question was, had our baby been in contact with the dog, and had the dog licked him? A course of injections could possibly kill the baby. We decided that the baby had not been in contact and then worried for the next twelve months in case he had.

Talikwa apiary

To study the African bees, I put swarms into Langstroth frame hives which initially I kept near my office which was close to the hospital. But beekeeping with the African bees is not compatible with other human activities, so I had to move them out into the bush about 10 miles from Tabora. Here at Talikwa, in an area of hills covered with miombo woodland gazetted as a Forest Reserve for Beekeeping Research, I established an apiary in a 'banda', a building having a roof but open walls, with only strong chicken wire to protect the hives from the honey badger or ratel. One of the hives I mounted on scales and I was able to observe and record all the different phases in the annual cycle of honeybee activity and to begin a study of the plants of importance to the bees. I found I had to make some alterations to the hives to suit local conditions and I started to use Modified Dadant hives and honey supers with Manley frames. I also made a study of the biology of the common African race of honeybee which showed some small differences from the European races.

I had to get to know the plants which provided bees with their food. In the vast open forest and woodland areas of the Central African Plateau, the trees clearly were of the utmost importance; most of them were leguminous. Much of my time on safari was spent studying the vegetation, collecting specimens for naming, recording times of flowering and observing bee activity. From the amount of rainfall I was able to forecast the approximate size of next year's beeswax crop.

I visited the markets and traders in all the villages in the beeswax producing area, met the chiefs and village headmen of the various native authorities, and got to know the leading beekeepers. From these last I recruited instructors to teach the best methods of collecting and rendering beeswax, using locally available materials. For these potential staff I held courses in beekeeping at Tabora, and took them out into the surrounding bush for practical work. One of the first beekeepers I recruited as an instructor was Saidi Salagata, a former Corporal in the King's African Rifles. I was still struggling to learn Swahili and Saidi was a strong right arm in maintaining good order and discipline.

Western Province, with its centre Tabora, was not the only major beekeeping area. The Southern Province was also important, exporting wax through the port of Lindi. The quantities produced were quite large but the quality was generally poor.

Saidi Salagata demonstrating extraction of beeswax

It took me a week to get there, working my way south through Western Province, across Southern Highlands Province, calling at all the centres of beekeeping activity on the way. From Njombe in Southern Highlands Province to Songea in Southern Province was by way of a tortuous track cut into the sides of the mountains, and across primitive bridges which needed careful inspection before crossing. The decking of one bridge, on the border between the two Provinces, was of bamboo. The Local Authority removed this before the rains were due and replaced it after the rains, effectively closing the road for six months of the year. This was not unusual; all the roads out of Tabora were liable to be impassable during the rains. By the time I had done a thorough tour of the beekeeping areas of the Southern Province and returned to Tabora, six weeks had elapsed.

Having a four-wheel drive vehicle did enable one to get home if caught out by a very wet spell. When Murray Lunan, the Provincial Agricultural Officer, saw my 1949 model Land Rover on its arrival, he exclaimed, "Now you will be able to rescue us". It was to be some years before this actually did happen.

The arrival of Stan Hubbard as my Assistant toward the end of my first tour meant that there would be someone to keep the work going while I was away on leave. After I returned to Tanganyika, I would be able to set up an out-station in Songea, which I had found to be the most important centre for the production of beeswax in Southern Province. Stan could then carry out extension work throughout that Province, supervise the African Beekeeping Instructors and collect data on the flowering of the nectar producing trees.

Stan had had an earlier career as a regular in the Royal Navy and was very capable and self reliant. Soon after he arrived in Songea, he was holding a meeting of the local beekeepers when he heard a swarm of bees coming. He dashed outside the baraza, picking up an empty 4-gallon kerosene tin and a piece of iron and proceeded to bang on the tin. This 'tanging' as it is called, was an old English custom for making a swarm settle. And it worked with the African bees. The swarm stooped its flight and settled close to the baraza. He then put the swarm into a hive. This demonstration established his reputation as a real fundi or expert.

It had been observed that cashew nut trees near hives of bees produced much larger crops than those more distant from the hives. As cashew nuts provide an important cash crop for farmers in the coastal areas of Southern Province, the Department of Agriculture was keen that beekeepers should be encouraged to place hives in cashew nut plantations and that the growers should welcome the beekeepers' hives or provide hives of their own. Stan Hubbard was busy explaining to the growers the benefits of having bees for pollinating the flowers. At one big meeting – this was all done in Swahili – a leading cashew nut grower complained that the flowers drooped after the bees had visited them. Stan came back quickly with the reply,

"You also droop after you have fertilised you wife."

For a moment there was a stunned silence, then applause and acclamation.

Earlier in this account I spoke of *General Orders* and the government's financial and stores procedures with which I had to become familiar on my arrival in Tanganyika. The correct way of doing things was laid down in *General Orders* and *Financial Orders*, from the addressing of letters right through to the various allowances which officers of various ranks could claim for travelling, hire of porters for foot safaris, hotel allowances and so forth. An officer took responsibility for the contents of any official letter he might write. That letter went out over his own name and designation. Although an officer might be authorised to sign on behalf of his superior, he made that clear in the letter and used his own name and designation.

Each officer was responsible for the expenditure of government funds allocated to him for his work. About half way through the financial year officers were required to make an estimate of their expenditure for the next year. This would have to cover salaries and wages of staff, travelling expenses and allowances, equipment and tools, purchase and running of government vehicles, maintenance and upkeep of stations etc. A week or two before the financial year began each officer received a warrant to expend the funds voted by Government and allocated to him by the Head of Department under the various sub-headings.

These details were carefully recorded in what was called 'The Vote Book'. Throughout the year, every item of expenditure was likewise recorded in the Vote Book, and each month the officer checked the expenditure shown in the Vote Book against a statement of account produced by Treasury. Originally the expenditure each officer made was identified by his designation and signature, but as central account keeping became more computerised, each officer also had a 'Vote Control Number' which he used on every voucher. Each officer was responsible for ensuring that the vote book was kept up to date and that he did not overspend the amount allocated. Also, if he did not spend the amount he had estimated and which had been allocated, he would be required to explain to Treasury through his Head of Department. The whole system seemed to work very well indeed and each officer was totally responsible for the correct expenditure of the funds allocated to him.

Stores were also carefully controlled. Tools and equipment purchased locally or received from Government Stores were signed for and recorded in the stores ledger and all issues similarly entered and the recipient signed an appropriate voucher. Each year the stores held by each officer were checked by a Board of Survey, consisting of two or three officers from other Department. This took place on the first working day after New Year's Day. Any discrepancies between the tools and equipment held by the officer and the details shown in the ledger had to be explained, either to the satisfaction of the Board or to the Head of Department or to the Government Auditor. The Board gave permission to write off breakages of laboratory glassware or other unavoidable losses and forwarded its report to the Head of Department.

Every officer had to write a monthly report and submit it to his immediate superior. I received simple but very informative reports from the African beekeeping instructors, written in Swahili, some in Arabic script. In turn I wrote my monthly report to my Head of Department. I found it an excellent exercise for organising one's thoughts and reviewing the work of the past month. I suppose that for me it was less of a chore because my work was interesting and varied, though I did not meet any Colonial Service officer who had a boring or unexciting job. The monthly reports made the building blocks for the all important Annual Report, and helped to make sure nothing of importance was overlooked.

I have made some mention of these routine matters because from what I have seen of some other civil services, the flaws of which rose to headline dimensions here in the 1980s, Governments could learn much from the Colonial Service, later known as Her Majesty's Overseas Civil Service.

My final, and really big achievement of that first tour was to get my research station built to my design in 1952. It contained two offices, one for me and one for my clerk, a microscope laboratory with storage for an herbarium, a chemical laboratory and a photographic darkroom. The other end of the building contained a big store room, a small workshop and a large honey processing room, which also served as a lecture room.

Beekeeping Research Station, Tabora

The building was made of granite, quarried from the kopje across the road, and it was situated on an area of regrowth 'miombo' bush, 250 metres square, alongside the road to the Tabora aerodrome. The Public Works Department built the research station building, but I, with the assistance of the Officer Messenger, Msema Kweli, and the Driver, Salum Cherehani, built the two-bay garage with a roomy inspection pit for the maintenance of our vehicles. In this area of bush I kept some hives of bees, a safe distance from the public, and in my office, close to my desk, I installed a Perspex observation hive. A water main was already in existence along the road to the aerodrome but I had to wait for two years for mains electricity; in the meantime I operated the centrifuge and provided lighting for the microscope, the darkroom enlarger and safelights with a portable 500 watt generator.

CHAPTER 6
BEE BOTANY

In the late spring of 1953, after three and a half years in Tanganyika, Joan and I and our four children went back to the UK on leave.

That first journey home from Tanganyika had its moments. We travelled by DC3 from Tabora to Nairobi where the airways put us into an hotel for the night, unfortunately not of the standard of the Norfolk. The kitchen staff could not even cook an egg properly for the children's tea. Then the Hermes aeroplane, which we boarded for London broke, down at Khartoum during the night and we all had to try and get some sleep lying on the desert sand. There appeared to be no transit passenger facilities. And finally, when we arrived at Joan's parents' home in Carshalton Beeches, we learned that the owner of the house in Devon which we had rented for our leave had cancelled the arrangement: her husband had just died. So we, with four children, had nowhere to go except to camp where we were until we could find alternative accommodation.

I rang up Alex Morice Wilson, a solicitor in Aberdeen whom I had assisted in a legal matter a few years before, and he came back to me with the offer of a house in Torphins owned by James Morrison, a school teacher whom I had helped by teaching his Cubs about trees when I was a forestry student. In Aberdeenshire a solicitor or advocate is also known as a Factor or a Man of Business. Joyfully we caught the first Aberdonian express train north.

'Monaltrie' was the usual type of house in Aberdeenshire, two windows downstairs and two dormer windows above them, and the door in the middle. It had a very nice garden at the bottom of which was 'the wee hoose' where Jaimey Morrison's Uncle Sandy and Auntie Belle lived. Next door was the Police Station occupied by Sandy the Policeman. He had two cells at the back of his house, one of which he used for putting up his mother-in-law and the other for giving stray hitch-hikers a bed for the night. Across the road was the war memorial, a large vertical piece of granite, and next to that was the village hospital where Rosemary and David had been born. Joan was great friends with the Matron, who had served in West Africa during the war.

We were rather short of funds, so we did not have a car and lived very quietly on that leave. After I had recovered from the journey, I set about writing a book on beekeeping. I intended it to be a fairly elementary book on basic beekeeping which would have a broad application, and could be of use to African beekeepers.

Time for summer holidays for the local people came round. Mr McKay, the Postmaster, who was also the Chemist and ran the lending library, asked me if I would care to relieve the postmen so that they could have their holidays. I was delighted; this would bring me in some very welcome cash. There were four postmen, and I accompanied each in turn on his round, was introduced to the farmers on the route, and learned at which farms the farmers' wives provided refreshments, tea and scones and cakes. They all liked 'to have a news'. For most of them the postie was their only contact with the outside world during the day. Pushing the heavy bicycle up the hills and along the rough farm tracks and footpaths was relieved by occasional freewheeling runs down sealed roads. It certainly made me fit and got the grey cells working. That went on for two months, and of course in all weathers.

Before being caught up in the postie business, I had called on Professor Steven, head of the Department of Forestry at Aberdeen, and had mentioned that I was interested in doing a Higher Degree. He referred me to a book by Dr G. Erdtman of Sweden. It had caught his attention and he thought it could give me some ideas. The book was *An Introduction to Pollen Analysis*, a subject which had not been explored in Africa and which had relevance to the beekeeping industry. If I carried out a research project n my own in Africa, I could eventually prepare and submit a thesis to the University for consideration for the award of the higher degree of Doctor of Science.

While plodding round the countryside, pushing the postie's bike, I gave much thought to this project and made notes of my ides as they were formed. Always I found that my best ideas came to me when walking in the countryside, or in later years even in suburban streets in the quiet of the early morning. By the time I had completed my postie duties, I had drawn up the outline of my research programme.

The first essential was to collect pollen grains from all the common flowering plants, trees, shrubs and herbs, and any other plants on which I found bees at work. At the same time I would collect specimens of leaves, flowers and fruit to send them to the East African Herbarium for naming, with enough spare material for mounting for my own reference collection.

The pollen which the bees stored in the combs, and which I could also collect from the legs of the bees themselves, would show me which plants were of importance as sources of pollen, the protein food of the bees. I could also extract pollen from honey, preferably directly from new honey comb. That would show which plants were important as sources of nectar.

I had to develop a technique for preparing the pollen grains for examination under the microscope, of describing them and of recording the descriptions in a manner that was readily retrievable. I needed to be able to compare identified pollen grains with the pollen collected by the bees or which I had extracted from honey. I had already begun a collection of herbarium specimens and pollen samples in 1951, and had experimented with various methods of mounting pollen grains on microscope slides. I tried the whole range of recommended stains, but nothing satisfied me until I used Erdtman's acetolysis method. This, although involved, produced clean specimens showing very clearly the structure of the wall of the grain. The yellow stain the grain took on was very suitable for photography in a pale blue light.

For recording the descriptions of pollen grains, and for the ready recovery of those descriptions, I selected the Paramount record card system and designed a Pollen Description Card. I used the characteristics of structure, shape and size defined by Dr Erdtman.

Pollen Description Card

This type of card has a hole near the edge opposite each feature, and the holes of the features present in the grain described are punched out so that the cards can be sorted by putting a long needle through the relevant holes: the cards of pollen grains having those features fall out.

I used the same system for recording plant specimens on a Botanical Record Card which had provision for sorting by collector's number, family, genus, vegetation type, flowering period and area and included particulars of habitat. A further refinement was an Ecological and Phenological Record Card which described the vegetation community, its habitat, locality and floristic composition, relevant details of each species and frequency of occurrence and month of flowering.

While all this had to wait until I returned to Tanganyika, the ideas were formed in the foothills of the Highlands during that very pleasant summer leave. Other notable events at the time were the ascent of Everest, the Coronation of Queen Elizabeth II, which we watched on Dr Morrison's television, and the Suez Canal crisis. In the village, the coronation was celebrated the presentation of coronation mugs to all children and by a children's sports day – ours gave a very good account of themselves.

The Royal Highland Agricultural Show was held near Edinburgh. I went with my old beekeeper friend, Alastair McCrae, in the company of the local farmers. Having already been involved with Agricultural Shows in Tanganyika, I was interested to see how it was done at the Highland. Also I had instructions from Joan to find out what I could about keeping chickens on deep litter so that we could keep them safe from predators. I was lucky, deep litter chicken houses were considered the right way to go, so I returned with lots of information which we put into practice on our return to Tabora.

Before the end of that leave, the Director of the Bee Research Association, Dr Eva Crane, who now lived near Gerrards Cross in Buckinghamshire, invited me to write a review on beekeeping in the tropics. I stayed with her for a while, reading up the references in the BRA library, and when I returned to Tanganyika I prepared the review for publication. It was published in *Bee World*, December 1953, under the title of 'Beekeeping in the Tropics'.

This time when we left Torphins, the whole village seemed to turn out to see us off, singing "Will ye no' come back again" as the train moved out of the station.

Our return flight to Tabora also had its moments. On arriving in Rome, we learned that fog in Cairo would delay us. Buses took all the passengers up to Roca di Papa in the hills above Rome, where we were put into hotels for the night; not that we slept much in the strange environment, especially with the local population apparently dashing about all night on Lambretta motor scooters. We were also entertained by the fiery confrontation of two Roman bus drivers whose buses had had a mild collision. We got away from Rome early the next morning; the fog in Cairo would have cleared by the time we arrived.

At Cairo, during the refueling stop, we were shepherded into the terminal building and to a lounge where a television was going full blast showing Nasser, the President of Egypt, taking continuously. I could not understand what he was saying but apparently it was anti British. The passengers conversed among themselves completely ignoring the television; "all being very British" as Joan put it. The end result was that we arrived at Nairobi very late, after the Customs and Immigration people had gone home to bed. We waited while they were rounded up to clear us from the airport and to let us go to our hotel for the night. It was not a very good return to East Africa. The officials were grumpy, the passengers disgruntled and the children overtired.

Back in Tabora we reoccupied our house which had been looked after in our absence by Stan Hubbard and his wife, and rapidly we settled back into the routine of life. Travelling all round Tanganyika visiting the beekeeping instructors, beekeepers and beeswax traders gave me the opportunity to get started on my research on the plants of importance to beekeeping. I also set up some small outstations in key areas with a hive of bees on scales. The local beekeeping instructor recorded the daily weight and observed and recorded the plants on which bees were working.

An interesting break in my routine came when I received from Dr Igor Mann of the Department of Veterinary Services in Kenya, an invitation to visit some areas where that Department was attempting to encourage beekeeping and the marketing of honey. Most valuable to me were the meetings I had with Mr Townley who kept bees in the Bahati Forest at 7,000 feet and in the Londiani Eucalyptus forests, and then with Mr Jim Nightingale at South Kinangop where he kept bees at 8,500 feet and also in the Rift Valley and Athi Plain. Both of these gentlemen were keeping bees on a commercial scale.

Mr P.J. Greenway and Mr B. Verdcourt of the East African Herbarium were now established in Nairobi so I visited them. I had met Mr Greenway a few years before when the herbarium was at the old German research station at Amani in the highlands of the Tanga Province of Tanganyika. These two botanists were to be of great help to me in my research, by identifying the plant material I sent them in a continuous flow over the next three years.

My laboratory equipment was simple but adequate for the work I had to do. I now had a good microscope with the best lenses for examining pollen grains, but lacked any camera equipment designed for photomicroscopy. I had already obtained a Rolleiflex twin-lens reflex camera with a 35 mm film adapter and plate holders. I made a wooden mount for the camera and experimented to find the most suitable exposures, using plates which I could develop after taking each picture. Having determined the right exposures for different sizes of grains, I set up the camera with the 35 mm adapter to do the routine photography; a primitive, Heath Robinson arrangement, but it worked.

When I was in Kenya in 1953 with Dr Mann there had been come concern about a terrorist organisation called Mau-Mau. The reaction in Tanganyika was to set up a Special Constabulary. I joined and almost immediately I was made Adjutant, looking after the administration of the Tabora Special Constabulary. When the Commandant, John Groome, who was also the Provincial Forest Officer, went on leave and was transferred in his work to another station, I was made Commandant, with the rank of Chief Inspector and with 350 officers, NCOs and men under my command.

We had a very enthusiastic force of Africans, some of whom had been in the army during the war in the King's African Rifles. The local tribe, the Wanyamwezi, had a long tradition of service in the army and the police – it was men of this tribe who served as porters for Livingstone, brought his body back from where he died in Northern Rhodesia and carried it all the way down to the coast. The Special

Constabulary Warrant Officer, Kinani Ali, was an ex KAR Sergeant-Major and had everything beautifully organised in the best British Army traditions. Happily we had no terrorist problems in Tanganyika and the Special Constabulary merely rendered assistance to the regular police when needed.

Because my work and that of my team of instructors was almost entirely confined to the forest areas, Government decided, at the suggestion of John Groome, that the Beekeeping Division, as it was called, should be transferred from the Department of Agriculture to the Forest Department. Three factors facilitated the transfer: the fact that I was a forestry graduate, the area in which my main research apiary was sited was a Forest Reserve specially declared for the purpose of beekeeping research, and the Member for Agriculture and Natural Resources – an officially appointed Minister whose telegraphic address was MANURE – had both the Department of Agriculture and Forests in his portfolio.

The Department of Agriculture had been most supportive of my work and in obtaining funds for staff, equipment and buildings and in advising and assisting me in the techniques of extension work among the producers. However, my work in the forests, and the interaction of forest beekeeping and normal forestry processes would be made easier by our being a part of the Forest Department. It transpired that in the long run it was a very happy arrangement. As far as the Beekeeping Instructors themselves were concerned, probably the only difference was that they were issued with a very nice line in green jerseys for wearing in the cold weather, but not suitable for wearing when working with bees.

To observe closely what went on in a hive of African bees, I had a small hive beside my office desk. This consisted of three perspex boxes, each of which held one frame of comb, the three combs being arranged one on top of the other. While I could observe what the bees were doing on the faces of each comb, they could not escape into the office, but had a Perspex tunnel leading from the bottom of the observation hive to a flight hole through the frame of the window behind my desk. Having an observation hive beside my desk tended to be a distraction, but with it I was able to observe the biological differences between African bees and European bees. The rate of development of both worker and queen larvae of the African bee was more rapid than that of other races.

I carried out experiments like those performed by Karl von Frisch in Austria in 1944. I observed and timed the dances of returning forager bees, which they perform to indicate to other bees in the hive where to find sources of nectar and pollen. I compared these dances with those which had been recorded as being performed by other races of honeybees.

During this tour of duty in Tanganyika much of my time was occupied with the administration of the expanded staff of Beekeeping Instructors and visiting them at their various stations throughout Tanganyika. My assistant, Stan Hubbard, was able to do great work in improving the quality of beeswax offered for sale in Southern Province, which was cut off from the rest of Tanganyika during the rains.

Although my findings on the adulteration of beeswax with Trigona was were published by the Royal Entomological Society in June 1954, there remained another two and half years of uphill struggle before Government agreed to my recommendations on the Beeswax Export rules.

Nevertheless, my research into the sources of nectar and pollen progressed and towards the end of that tour I assembled my data and wrote my thesis. Having the pollen descriptions on the Paramount record cards, and botanical collections similarly recorded, was a great help. Today one would probably use a computer, undoubtedly faster for retrieving information, but vastly more expensive and less portable. But would it have enabled me to do more work and produce better results? I doubt it. The real help would have been in the writing up of the results, which task has now been revolutionised by the personal computer word processor. A major task was printing the photographs of pollen grains to the most acceptable standard, and mounting and labelling them correctly. Eventually, five copies of each of the two volumes of the thesis, text in one volume and photographs in the other, were bound by the White Fathers at the Mission Press at Kipalapala, a few miles south of Tabora.

Before I went on leave in 1956, the Chief Conservator of Forests in Kenya asked for me to visit forest areas there to examine the potentialities for beekeeping development. The Mau-Mau emergency had made it necessary to concentrate forest workers into villages under the protection and control of the Forest Department. This presented valuable conditions and facilities for the development of beekeeping as a profitable side line for the forest workers. Under the guidance of the Forest Department, this tour enabled me to visit areas of Kenya I had not seen before and to go to areas normally inaccessible because of the emergency.

The actual journey back to the UK early in the spring of 1956 was uneventful. This time we had succeeded in saving up enough to buy a caravan and a new Land Rover, both of which would be awaiting us on our arrival in England. The Land Rover had to be picked up from a Rover depot in London the day after our arrival so, at the airways terminal at Victoria, we asked if they could recommend a convenient hotel for the night, and we were delighted with the advice we were given.

After breakfast I picked up the vehicle while Joan organised the four children. This was my first experience of driving in London, though I had done a lot of cycling through London before the war, but I found that, so long as I knew well in advance, just where I was going and which turns to make, it was not too bad. The essential was a clear street map. Once on the main roads, the direction signs were excellent, and the speed limit on each particular stretch of road was shown repeatedly. So, with the new Land Rover loaded with the six of us and our baggage, we set off towards Southampton to collect our caravan.

The van we picked up at Pennant Caravans was 22 feet long. It had a room at the rear with four bunks for the children. The central living room had a double bed which folded away into one wall and at the other end was a solid fuel stove which also heated the water. At the front end there was a neat little galley on one side and on the other a bathroom complete with wash basin, minute bath and chemical toilet. Cooking and lighting were by bottled gas so we were completely independent of electricity supplies.

Having equipped ourselves at Pennant Caravans with Melamine plates and mugs we set off, first to call on Joan's parents who were now at Worthing, and to show off our new acquisitions, and then to go down the road to camp discreetly at the edge of a field concealed by the trees alongside an ancient road north of Angmering.

Once we were into our mobile home, my first task was to apply to the University of Aberdeen to be admitted as a candidate for the Degree of Doctor of Science and to send off to the University two copies of my thesis, *Bee Botany in Tanganyika*.

As Easter was approaching, the local caravan parks were fully booked. We did not want to be on the road at that busy time, so we called in at the Forest Office in Lyndhurst and obtained a permit to camp in the New Forest. We had selected a likely spot from the ordnance survey map. We were quite on our own, near a very small stream. When we made camp, the trees were still leafless, but over that Easter weekend, the weather was beautiful and the trees unfurled their leaves and by the lime we left they were covered in their fresh spring foliage. It was an enchanting time. We emerged from the forest only to go to church on Easter Sunday morning in Lyndhurst.

Easter over, and the holiday traffic clear of the roads, we went down to Salcombe in Devon where we made camp on a steep hillside on a farm, which had several other caravans staying there for fairly long periods. Our purpose in going to Salcombe was to do some sailing. We had joined the Island Cruising Club which seemed the ideal club for our needs. It was a club for members who do not own their own boats but who sailed in the club's boats, everything from 13-foot dinghies to a 75-ton Brixham Trawler, *Provident*. All kitted out with woolly caps and life jackets, as a family we sailed in the 13-footers, and Christopher spent a lot of time enjoying himself rowing a tiny clinker-built pram dinghy, aptly named *Flying Saucer*. To keep myself calm while waiting to hear from Aberdeen, I worked as a bosun on a beautiful old schooner, *Hoshi*. Antony was still too small to venture afloat in his own but he made friends with a small boy of about his own age in a nearby caravan, the son of Angus Primrose, the yacht designer, who at that time was Secretary of the ICC.

Then came the wonderful news, first a letter from Professor Steven and then a letter from the Secretary's office to the effect that my thesis had been sustained and that the degree of Doctor of Science was to be conferred upon me. The Professor told me that I was the first of the post war students to have attained it and that the number who receive it at Aberdeen is distinctly limited. Joan and I went down to the 'Fortescue', the pub patronised by the ICC, and celebrated with other members, many of whom were university students and appreciated the significance of the event.

The letters from Aberdeen had been written on 26th June and it was necessary for me to present myself at Marischal College for enrolment before the 4th July, Graduation Day. We prepared to move the next day, but before we got away David had to have an emergency visit to the dentist, late in the evening.

Our journey from Salcome to Aberdeen towing the caravan with the Land Rover ballasted with the children, gas bottle and sack of anthracite for the stove, and with speed limited to 30 miles per hour, took us four days. We stopped for the night at any convenient spot, once in an orchard just off the road, another time on a wet miserable evening in the north of England in a road maintenance area, next to the snowplough. We passed through Wigan in the rush of a Saturday morning and we ground our way over the passes in the border country in company with the heavy transport. On arrival at Aberdeen we made for Hazelhead where the Council ran a caravan park for visitors on well kept lawns among the trees.

Neither Joan nor I was equipped for the functions associated with the graduation. I got myself a suit off the peg in Aberdeen and Joan's friend, the Matron at the Torphins hospital, lent her a glamorous gown for the reception and ball. I hired academic gear from the resources of Esselmont and Mackintosh. We attended the formal graduation ceremony on Wednesday the 4th July, followed by a lunch given by the University on the next day and the Graduation Reception in the evening. We were very fortunate to be camped so close at hand at Hazelhead.

It was also fun for the children. The trams ran from the terminus at Hazelhead, down Union Street into Aberdeen and then out to the Brig of Don in Old Aberdeen. The children, who had never seen a tram before, would do the whole trip to the Brig of Don and back on the upper deck of the tram, repeatedly.

When all that was over we took ourselves off to Torphins where we camped for several weeks in a convenient field. The farmer, who supplied the area with milk, entertained us with a great fund of hilarious agricultural stories.

In my capacity as Commandant of the Tabora Special Constabulary I introduced myself to the Aberdeen Police. They were most helpful and showed me how their system of patrolling worked. Because of the shortage of policemen after the war, they developed what became known as 'The Aberdeen System' which was adopted by other police forces with similar problems. This system was based on the recording of all complaints, crimes and accidents on Paramount punched cards. From this information, where and when most incidents occurred could be determined. The police concentrated their patrols on those particular areas at the appropriate times. I was taken out in the patrol cars to observe the system in action.

While we were at Salcombe, the children had become familiar with the books of Arthur Ransom which Joan read to them. So, on our return journey from Scotland we visited the Lake District and the scenes of *Swallows and Amazons* and other stories, and identified some of the features around the lakes mentioned in the books. The rest of that leave we spent partly at Ferring close to Joan's parents and to my Uncle Lionel and Auntie Maud at nearby Worthing, and partly at Dawlish where my mother lived.

One important task that had to be done before returning to Tanganyika was to get new calling cards printed with my recently acquired title of Doctor. It was still the custom when returning from leave, or going to a new station, to leave cards on the Provincial Commissioner, the District Commissioner, the Senior Medical Officer and anyone else rather higher than oneself in the pecking order.

We returned the caravan to Pennant Caravans for storage until our next leave and then we took the Land Rover back to the depot in London for forwarding on to us in Tanganyika. Until that arrived I would be depending on the government Land Rover for transport.

As I cannot recall anything about the flight back to Tabora, it must have gone smoothly, at least until we landed at Tabora. When I came to the door of the DC3 and looked across to where people were waiting to meet passengers off the plane, I saw my Beekeeping Division Land Rover. Clearly it had a broken front spring. Somehow we got home. The broken spring was replaced in record time.

Now, with the marathon effort of my study of bee botany behind me and armed with the diploma of the degree of Doctor of Science, I felt ready to face any new challenges which came my way.

CHAPTER 7
COMPLETED PROJECTS

On my return to Tabora in October 1956, I found that two big events were about to take place. First the visit of Princess Margaret to Tabora concerned mainly the Administration, but the Police were deeply involved. The role of the Special Constabulary had been provided for by Peter Tozer, my very able adjutant, so I had nothing to do other than to attend the official reception in uniform.

I took along with me my own personal visitor, Dr Warwick Estevam Kerr of Sao Paulo, Brazil, who had just arrived via South Africa. He had came to get some of the vigorous, highly productive, African bees and to see the stingless bees that made double-sided vertical brood combs. Dr Kerr owes his name to a Scottish engineer grandfather who went to Brazil to build railways. He had read my 1954 paper describing the nests and waxes of Tanganyika's four larger species of stingless bees and, being a geneticist at the University of Sao Paulo and greatly interested in bees, he had to come and see them for himself and to count their chromosomes.

African bees for Brazil

About the time I had returned from leave, the hives of bees at the research station on Aerodrome Road had been attacked by a honey badger, otherwise known as a ratel. This animal is so powerful that blows from its claws make hives look as if they have been smashed up by machine-gun bullets. Of course the bees were furious and uncontrollable. I had to make a number of sorties in order to reassemble the remains of the hives. I mounted guard at night myself in the hope of catching the animal at his work of destruction. He either sensed my presence or had satiated his appetite for honey; I never saw him or did he attack there again. But it was clear that I would have to do something about this apiary because even occasional honey badger attacks would wreck my experiments and make hive management impossible.

So when Dr Kerr said that he wanted to send some of our Tabora queen bees to Brazil, I told him that if he could find the queens in the hives near the research station he was welcome to them; they were, I knew, of good honey producing stock. So, wearing my protective clothing, Warwick Kerr got to work. Searching for queens on the combs of African bees which will not be subdued with smoke is quite a nerve wracking business; they just keep on attacking. But Warwick is a very brave man; he got his queens, sent them off to Brazil by airmail and they arrived safely. Later, after some experience of them in Brazil, he described them as "the most vigorous and productive bees he had ever encountered".

That part of his work done, I took him to Ilonga to look for the thin bodied Trigona bees which make nests in the open and have double-sided vertical brood combs. After some searching in the bush we found a convenient nest under the branch of a fallen tree. We removed the nest and took it back to Tabora in a cloth bag so that Warwick had plenty of material from which to select the stages at which chromosome counts could be made.

I think that the visit of a scientist of the calibre of Dr Kerr was most reassuring to me. For seven years I had been working on my own with no yardstick against which I could measure my progress or the worth of my work, and there was no one working in the same field with whom I could discuss matters face to face. The Bee Research Association was growing under the directorship of Eva Crane producing *Apicultural Abstracts*, the *Journal of Apicultural Research* and the long established *Bee World*. The BRA was my only back-up facility. It was a lonely path I was treading and Warwick Kerr's visit was a great help.

Government still had not accepted my recommendations concerning the Beeswax Export Rules. I had done all I could about the improvement of beeswax; I had done the necessary research, made my recommendations to the Head of Department including drafting the new regulations, and had trained and put in the field a team of Instructors; it was up to Government to take the necessary legislative action to bring the adulteration of beeswax to a full stop.

During this frustrating period, I had turned my attention to two other aspects which needed attention, honey marketing and the management of the highly productive but intractable African honey bees.

Honey marketing

The problem with the honey was that large quantities were being produced away out in the bush, but there was no way that the beekeepers could bring it all in to market. In their bush camps, often on the sites of villages from which they had been removed many years before on account of sleeping sickness, the beekeepers separated the honey from the wax which they then rendered into clean cakes. These cakes of wax, together with a little honey for their own use, they carried on their heads from the bush to their new settlements. They sold the wax for cash in the markets and the honey they either used as food or made into beer.

It was the honey which was being dumped in the bush which was such a tragic waste. Much of this honey, as produced by the bees, was absolutely first class and could compete with any on the world market, if it could be got there economically and without deterioration in quality. For that to be achieved there were difficulties to be overcome. The first was to separate the honey from the wax in such a manner that the quality of the honey would not be damaged. The traditional method had been to heat all the combs up in a large pot and skim off the wax from the surface. The heat, and the contact of the honey with brood comb and pollen, caused a severe deterioration in the honey quality, making it unacceptable on the world market and fit only for use in beer making.

The next problem was to supply the beekeepers in their camps deep in the bush with suitable containers into which to pack the extracted honey, and then to transport the honey back to the central plant where the final preparation and packing could be done.

The third problem was to devise a simple plant which would refine the honey, without damage, to a standard acceptable on the world market. One thing here was certain, any process involving heat had to be avoided: the ambient temperatures caused browning over a period of time without this being accelerated by high temperatures during processing.

Then there was the need to pack the honey in containers acceptable to the market, and to protect those containers from damage in transit. Finally there was the need to store honey awaiting shipment in such a manner that the ambient temperature would not cause continuing deterioration of colour and flavour. Compared with honey, beeswax was a very easy and stable article of trade.

One of the earliest matters tackled was that of packing materials. The Metal Box Company in Dar es Salaam was able to supply 4-gallon tins made of excellent quality tinplate, lined internally with acid resistant lacquer and with lever lids of about 100mm diameter. These tins were filled with 25kg of honey, leaving about 25mm air space to absorb the shocks of travelling. As it was important to protect the tins from damage, I obtained several makes of corrugated cardboard cases for testing. Remembering what I had seen of the handling of goods at the Southern Railway stations in London, I tested a number of cases of each make, each containing a tin filled with honey. The test was quite simple; I threw each packed tin off the back of a lorry into a concrete inspection pit which had a concrete bottom, a drop of three metres. The make of case which protected the honey tins for the greatest number of drops was the winner.

The honey as collected by the beekeepers was in combs which they had cut out of simple hives. The only way to separate the honey from the wax without heat was to press the comb. The best press for the job was the English-made Mountain Grey Heather Honey Press, manufactured in steel and hot-dipped galvanized iron. Once I had demonstrated these presses, individual beekeepers and embryonic produce-marketing co-operatives were keen to buy them. But because of the high cost and the time it took to get supplies from the UK, I designed a simple lever honey press which I had made locally.

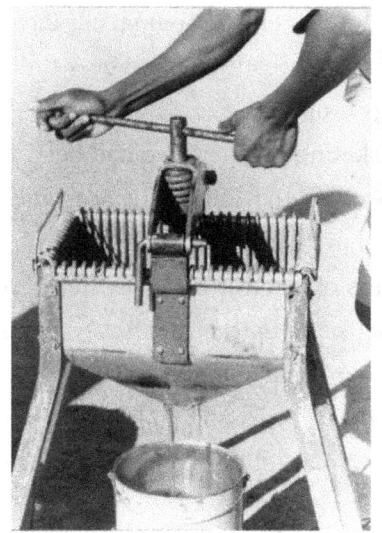 

Mountain Grey Honey Press   *Lever Honey Press*

To get trucks to the beekeepers' camps, the beekeepers themselves and their families cut tracks through the bush. These were rough and unsurfaced, but as the honey was not ready for harvesting until well into the dry season, the tracks were passable, and they improved with use and a little maintenance. They were, however, hard on vehicles, especially when loaded with honey. In the earliest stages of this work we used a long wheel-base Land Rover 107 and frequently suffered a bent back axle, with consequent failure of wheel bearings. Rover designed and made a reinforced axle casing and improved wheel bearings, which I fitted, and the failures ceased.

The next stage was to set up a central plant to strain the pressed honey and so remove the small particles of wax which passed through the cloth in the presses. Initially I tried the Ontario Agricultural

College honey strainer, but this had been designed for handling honey from centrifugal extractors used by frame-hive beekeepers. The filter screens clogged too rapidly for it to be practicable in our situation. So I designed a tank in which the honey flowed over and under a series of baffles which held back the wax particles as they floated to the top of the honey. From the bottom of the last compartment the almost clear honey was pumped up into the OA strainer for the final clarification. Once I had mastered this problem I was ready to hand over the whole operation to the honey-marketing co-operative.

A very successful produce-marketing co-operative had been established among the coffee growers of the Chagga tribe on the slopes of Kilimanjaro. The Kilimanjaro Native Coffee Union was the model which I applied to the groups of beekeepers in Western Province who were keen to organise themselves to market the honey which was otherwise being wasted. After two or three years of exploring the market and developing the techniques, I was able to hand over the whole organisation of collecting and marketing the honey to the Co-operatives.

Export of beeswax

On the 1st February 1957, the *Produce Export (Beeswax) Rules, 1957* were published under the Produce Export Ordinance. I went to Dar es Salaam and showed the Customs Officers the technique of examining consignments of beeswax and how to recognise good beeswax. They were to take samples of anything which appeared doubtful and to send them to the Government Chemist for immediate analysis, in the meantime holding the consignment. It was necessary to prosecute only one exporter for making to Customs a false declaration about the quality of the beeswax. Conviction resulted in confiscation of the consignment and a fine of three times its value. Henceforth traders refused to buy adulterated beeswax and the whole trade in that product was cleaned up. But it had taken me seven years of fighting to get there.

Beeswax film

I wrote the script for a film on the production and marketing of Tanganyika beeswax and, having obtained a compact 16mm movie camera, set about making the film. After editing it and providing captions, I sent it to Colour Film Services in England for copies to be printed from the Kodachrome original. There were several 16mm projectors in the Beekeeping Division so we were able to show the film as a teaching aid to beekeepers in the bush. I entered a copy under the title of *Tanganyika Beeswax* for the Apicultural Congress in Rome in September 1958 and won a silver medal in the documentary section.

Bee houses

I was now able to concentrate on the problems of hive and apiary management. With simple log or bark hives, management was confined to baiting the hives was wax or propolis to attract swarms of bees and to siting the hives high in trees where they would be safe from ants and the honey badger. Such hives could be visited only once a year to collect the honey crop, often depriving the bees of the food they needed to survive the dry season and causing them to abandon the hives and form hunger swarms, many of which died of starvation.

Generally the hives were widely scattered in the bush, so the beekeeper spent much of his time walking from hive to hive, climbing trees, lighting up his bundle of twigs or grass to make smoke and lowering the crop to the ground. Because of the viciousness of the bees, most of the work had to be done at night.

What was required was a beekeeping system which would enable the beekeeper to collect the crop from a large number of hives quickly, in daylight and without much danger of being severely stung. The hives needed to be protected from thieves, the honey badger and from ants, and to be grouped together in sufficient number to make the provision of supply of water practicable. It was desirable that some management of the bee colonies be possible to stock hives other than by the chance occupation by swarms and to maintain and increase the productivity of individual hives.

After experimenting with keeping the African bees in frame hives in an open sided banda, as well as on stands in the open, I decided to try keeping them in a bee-house. Bee-houses are used in parts of Europe, mainly for protection from the winter weather. The hives are arranged along the walls. Each hive has its entrance in close contact with a hole through the wall. The bees fly in and out of the hives through the walls without entering the room itself. The house is constructed so that it is completely bee tight but the windows are fitted with a bee-escape arrangement so that any bees which do get into the room can escape to the outside.

A Bee House at Tabora

I built my first bee-house on the research station site on the aerodrome road. It had a concrete floor and walls of cement blocks up to 120cm above the floor. Above that, 90cm high window frames were fitted with expanded metal to keep out thieves and with mosquito gauze to keep out the bees. A gap at the top of the wire with a hardboard baffle on the inside formed an escape for any bees flying in the room. As I was going to use frame hives for my research, these stood in pairs on stands at a convenient height for working, with their entrances in contact with a piece of timber secured to the wall. A hole in this timber led to a length of pipe fitted in the wall and projecting on the outside about 45cm. This was needed to give the bees the chance to defend themselves against driver ants. To assist the bees to identify their own entrance, I painted different patterns on the walls, and used colours which the bees could see as being different.

This research bee-house was quickly stocked with bees and results came up to expectations. It was at last a pleasure to work with African bees, even when the colonies had built up to full strength. I was able to do a demonstration for the Chief Conservator of Forests in which none of us found it necessary to wear any protection whatsoever. The bees in the hives were controlled with smoke in the normal way, and the guard bees, who cause all the trouble, just flew around outside, unable to get at us in the bee house. And because they could not sing anybody, they did not arouse the other bees in the hive.

Beekeeping with African bees now became a pleasure and the bees themselves thrived. I was able to use the more complex methods of hive management normal to bee farming in Europe and America, including the raising of queens from selected stock. I was also able to conduct proper courses in beekeeping for my more senior staff and give them confidence in the handling of bees.

After the success of the first bee-house at the Research Station, I built others out in the bush, teaching my staff how to build, using concrete, cement building blocks and corrugated iron roofing on timber trusses and purlins; I learnt later that the staff had put the knowledge obtained to good use, building themselves new houses of these materials, replacing the traditional mud and wattle and thatch houses which had only a very limited life.

A tree bee-house in Songea

While the bee-houses for the research apiaries were made of durable materials, others were being constructed by beekeeping instructors in many parts of the country, and also at schools which had beekeeping in their syllabus, using the materials for building which were ready at hand. These contained simple hives, sometimes in two or three layers. As the advantages of this method of beekeeping were seen by the beekeepers, bee-houses began to be adopted by the more progressive honey and beeswax producers. Some combined the old methods with the new by actually building bee houses up in the branches of trees.

Simple hives

I encouraged the continued use of simple hives for honey and beeswax production by the average African beekeeper because I could not see any justification for the great expense of modern frame hives under the condition in the main beekeeping areas of Tanganyika, other than for research, teaching, stock selection and queen breeding. I was, however, concerned that alternative materials be used for the making of hives; the traditional method of making hives destroyed the very trees which produced the best honey. The bark taken from a well grown and vigorous 'mua', *Julbernardia globiflora*, made only one hive and killed the tree.

If hives of the simply type could be made from sawn timber, many more could be constructed from a single tree. But this was a concept far removed from tradition, and would require much more work or the expenditure of cash. I had my doubts as to whether this would catch on in the traditional beekeeping areas of the 'miombo' forest. So long as there was a seemingly limitless supply of trees in the forest which covered half the land area of Tanganyika, the beekeepers would not see the point of it. And, when they realised that the trees suitable for bar-hive making were getting difficult to find, the plants which produce the nectar to support the bees would have gone and beekeeping would no longer be worth while.

Pollination

Originally I did not expect to find any pollination problems in Tanganyika because of the high density of the honeybee population and the abundance of stingless bees and other insects. However, the Overseas Food Corporation had cleared vast areas of woodland of all natural vegetation and so of bees. At Kongwa in Central Province, one third of the sunflower crop was lost through lack of pollination. The OFC employed an English commercial beekeeper to solve the problem, but before he had a chance to build up the necessary apiaries, they abandoned the growing of sunflower and took up ranching.

In a plantation of cashew nut trees in Southern Province, Agricultural Officers had noted that within 130 metres of a honeybee colony the trees were fully laden with fruit, but that further away from the hive the crop decreased. This stimulated cashew growers to take an interest in beekeeping and, from the vigorous and colourful extension work of Stan Hubbard, the need for bees in the plantations became widely recognised. Nearby, on the coast, an Agricultural Office made a similar observation in a coconut plantation.

Reports from coffee growers on Kilimanjaro suggested a relationship between the population of honeybees and the yield of coffee beans. The coffee growers were encouraged to keep bees for pollination, s well as for the fine quality of honey obtainable in the coffee areas.

Pyrethrum is grown commercially for insecticide. The flowers are collected when they are in full bloom and not later when the seeds have developed, so it seemed unlikely that pollination would be a factor influencing the proportion of pyrethrins in the flowers. Colonel L.A. Notcutt at Uwemba in Southern Highlands Province was interested in checking the effects of pollination of the yield of pyrethrins. Preliminary trials in 1957, using caged sample plots, showed that the yield of pyrethrins was greatest from the cages which contained hives of bees and least in the insect-free cages. The yield in the uncaged plots was intermediate. Feeding with sugar syrup scented with pyrethrum flowers was used to counteract the attraction of other plants for the bees.

Northern Rhodesia

It was in the dry season of 1959 that I was invited by the Forest Department of Northern Rhodesia to visit them and tour their main beekeeping areas and make recommendations for the development of the beeswax and honey production. Much of the country was 'miombo' forest, similar to that in Tanganyika. *Brachystegia spiciformis*, a common tree in the 'miombo', produced, in the copper belt, a honey light in colour and of excellent flavour. In Tanganyika it produced a dark coloured, strongly flavoured honey. Beekeepers using frame hives observed too, that where the soil was more fertile, the 'miombo' produced much better honey crops. The tour was enlivened by my being driven, while in the Rhodesian Teak Forest near the Zambezi, right under the trunk of a startled elephant, which was standing on a mound beside the track; the driver never saw it. I was also fortunate to be able to visit the Victoria Falls and to show my film, *Tanganyika Beeswax*, to a large audience at the hotel.

The report I wrote on *Beekeeping in Northern Rhodesia; Its Prospects and Recommendations for its Development* was, I considered, one of my better bits of work, so I submitted it as a thesis for the National Diploma in Beekeeping. It was accepted and I was able to add NDB to my qualifications.

Book on beekeeping

At the beginning of 1958, the Colonial Office invited me to write a book on beekeeping in tropical countries for publication by Longmans Green in the Tropical Agricultural Series. It happened that at that time I was in the act of preparing the lectures of an advanced course for beekeeping instructors, so I combined both tasks, to the ultimate benefit of my students as well as the clarity of the text of the book.

I had completed the book by the time I went on leave in August 1959 and took the manuscript with me. This time I arranged to hire a Land Rover for towing the caravan around England. After some delay, owing to confusion about the date of our arrival in England, the Land Rover was delivered to me at London Airport. We loaded our baggage and ourselves on board and set off on the road towards Southampton where we had left the caravan in store. After a while there were complaints from the children in the back because the baggage had shifted. I stopped to reorganise the load and then realised that I could not see my brief case containing the BOOK, nor my portable typewriter. They were last seen by the children on the edge of the pavement at London Airport. We returned anxiously and arrived at our place of departure. There were now many people standing on the pavement awaiting transport and I could see nothing of my brief case nor of my typewriter. I got out of the Land Rover, walked through the crowd and there, at the back of the pavement against the wall, placed neatly side by side, were my brief case and typewriter.

We picked up the caravan and made our way to Ferring to park near Joan's parents and my uncle and aunt. Once more the family had to be fitted out with clothes, this time to withstand a winter in England. I got myself a warm suit in Littlehampton and went up with my book to London to hand it over to Longmans Green in Clifford Street. It rained and I got very wet. I delivered the book safely but my new winter suit shrank.

In Worthing, on the strength of the pending publication of my book, I traded in my trusty ten-year-old Empire portable typewriter for a beautiful new Bluebird portable, which served me well for the next twenty-five years, until I succumbed to the convenience of an electronic typewriter. *Beekeeping in the Tropics* was published in May 1960, soon after I had returned to Tanganyika from leave. Although it went out of print eleven years later having sold 3300 copies, I found, when visiting Thailand in 1983, it was still in demand.

CHAPTER 8
LAST TOUR

Our leave in England extended from the beginning of August 1959 into mid-winter and we found that we were caught up in the programmes of meetings of beekeepers' organisations. I was asked to give talks on beekeeping in East Africa and to show my film, *Tanganyika Beeswax*. One such occasion, and possible the most august, was a combined meeting of the Central Association of Beekeepers and the Bee Research Association at the Zoological Gardens in London. The Central Association published the lecture under the title of *Beekeeping in the Tropics of Africa*.

Other organizations to which I lectured were the National Honey Show where I also took part in a beeswax symposium, the Cambridge Beekeepers Association's advanced course, the Reading Beekeepers' Association and Berkshire Federation of Beekeepers, the Slough Beekeepers' Association, the Twickenham and Thames Valley Beekeepers' Association, the Imperial College of Science and Technology at Sunninghill, the Chichester Beekeepers' Association and the Brighton Beekeepers' Association.

I also attended several conferences at the invitation of the organisers: the Honey Producers' Association Annual General Meeting, the Joint Conference of County Beekeeping Instructors and National Agricultural Advisory Service Officers at Harpenden and at the Ministry of Agriculture, Fisheries and Food and two of the Council Meetings of the Bee Research Association. I found all these occasions useful and instructive, giving me the opportunity of meeting and hearing all the leading figures in beekeeping in England and Wales. The Bee Research Association appointed me Honorary Secretary of the Association's African Committee.

Moving around with our caravan, I was able to meet beekeepers and bee appliance manufacturers, visit the Bee Research Department at Rothampstead, the Natural History Museum, the Commonwealth Institute of Entomology, the Tropical Products Institute , and to observe the work of the County Beekeeping Instructors. I also called on Mr D. Rhind at the Colonial Office; it was he who had approached me to write *Beekeeping in the Tropics*.

I was invited to attend at the Crown Agents where I interviewed Bill Reynolds, the successful candidate for the position of a second Assistant Beeswax Officer. Joan talked to his wife to make quite sure that she understood the conditions under which she would be living in Tanganyika.

Arrangements were made for the recognition of the system of beekeeping examinations I had instituted in Tanganyika. The highest level of examination, Expert Beemaster, was accepted as qualifying candidates for entry to the examination for the National Diploma in Beekeeping.

We remained in the south of England for this leave, moving from Ferring down to Dawlish Warren so that my mother could see the family. While there, and still having summer weather, I painted the caravan and got sunburnt. The only other time I was badly sunburnt had been at the age of ten, also at Dawlish. Since then I had served four and a half years in the Middle East Forces during the war and ten years in Tanganyika without being burnt. I always made a point of wearing my shirt.

Moving up to Cambridge and into East Anglia, we did some touring before going to stay at Denham, close to the Bee Research Association headquarters at Gerrards Cross. For Christmas we were back in Sussex, where the weather turned very cold. Antony, our youngest child, went down with malaria and while he was confined to his bed, we had a good fall of snow, the first the children had experienced since they were babies in Torphins. Naturally they made the most of it and as Antony could not join them, they made a snowball and took it into him so that he could touch snow for the first time.

*   *   *

I felt that I had gone as far as I could with research and development at Tabora and I now had staff who could carry on the work in the main beekeeping areas of Western and Southern Provinces. I knew little of the Northern Province where excellent honey was being produced on Kilimanjaro. There are great differences in the vegetation at different altitudes, so there was the possibility of a potential for migratory beekeeping. Also, the local Wachagga are a progressive and practical people who could become interested in modern commercial beekeeping, if I could show that it was worth while. So I decided that I would be most usefully employed examining the beekeeping potential on Kilimanjaro and to do this I would have to transfer my headquarters to Arusha, the Provincial Capital. Arusha was situated on the lower slopes of Mount Meru, a dormant volcano which rises to nearly 15,000 feet (4565 metres) above sea level.

On my return to Tabora, I sought permission to make the move and in Arusha, asked the District Commissioner about the availability of housing and a suitable building to use as an office and laboratory. The latter was the easier. The Provincial Education Officer was moving into new offices and his old building would be available. Here I was able to have the office space I needed, a laboratory and a darkroom. A nearby out-building served as a workshop and store. There was plenty of open space around, but as the office was in the town, I would not be able to have an apiary in the vicinity. But that did not matter; I was going to establish my apiaries at different altitudes on Kilimanjaro. Also, there was an Agricultural Research Station at Tengeru, not far from Arusha, where I could set up another apiary for research, queen raising and teaching.

Getting a house was not so easy. The housing committee allocated us accommodation in a block of flats which we did not like, so as soon as I heard that an officer was vacating his house next door to that of the Provincial Forest Officer, I applied for it and got it. I had stayed in that house when I took part in an Agricultural Show at Arusha in the early 50s when it was occupied by Arthur and Pam Hamersley. Before the post war development of Arusha, this house had been the home of a coffee farmer. It had a beautiful well developed garden, irrigated by a leat, and had several large trees, remnants of the original forest, much of which still covered the mountain. The Temi River, which supplied water for the town and our leat, passed through a wide but step sided gorge nearby. My office was on the other side of the gorge near the Police Station and the Native Authority offices, and I could walk to it from the house over a wooden foot bridge.

As we were at 4500 feet (between 1300 and 1400 metres) above sea level the climate was pleasantly cool and the water which came down the Temi from the mountain was a constant 20°C – ideal for processing photographs. The air that rolled down the mountain in the evening was cool enough to require the wearing of a jersey or jacket and in the sitting room we had a big open fireplace which we used in the evenings for most of the year. During the rains we had quite a lot of mist in the mornings and the climate was decidedly damp, so much so that the oldest part of my office, which had a concrete floor, grew fungus on the floor. A row of mighty fig trees behind the office contributed to the damp because they shaded the office all day long. It was with some reluctance that I had to arrange for the Public Words Department to cut them down.

Another advantage in living in Arusha was that it was fairly near the border with Kenya and was at the start of the main road to Nairobi. Two of the boys, Christopher and David, were now going to school at St. Mary's in Nairobi; Antony was at school at Soni in the Usambara Mountains towards Tanga. It was only an 80 kilometres drive on a bitumen road to the railhead of the Tanga line at Moshi.

Rosemary had to be taught at home; there was not yet a vacancy for her at Loreto in Nairobi. So Joan taught her and before long she had two more pupils, one the daughter of a settler who in exchange taught Rosemary riding, and the other the daughter of a missionary who taught Rosemary the piano, a convenient arrangement for them all until they were able to get into their schools in Kenya.

Light bee-houses

Following my earlier success with keeping hives in bee houses, I built one at Tengeru; this time I tried out a light construction which I could use for the apiaries on Kilimanjaro. I made concrete stumps which I set in the ground to support the main bearers of timber. On these I laid the floor joists and then the tongued and grooved floor boards, the whole being well treated with creosote. The walls were of timber frames, bolted together, and clad with tempered hardboard up to 180 centimetres and topped by 60cm of expanded metal and mosquito gauze to form the window and bee escape. The ceiling was also tempered hardboard to make the room completely bee proof. The roof was made curved like a railway carriage but with wide eves and covered with tempered hardboard. The whole was well sealed with several layers of paint. The door was of normal timber construction.

I brought hives stocked with bees all the way from Tabora; not without casualties from panic and overheating. I learned a few lessons about transporting bees on the trip but I never tried the Australian method of moving bees with open entrances. With African bees it could be an interesting experience; one might learn some very tough lessons!

Once the apiary was well established at Tengeru, I started building the bee houses on Kilimanjaro. The first was at about 1800 metres, in high forest close to the road from Marangu to Bismark Hut. This bee house was five metres long with stands for eight hives at one end and at the other end, nearest the door, space for storage and for camping. Then at 2400 metres, in Podocarpus forest with a very high rainfall and about 300 metres below Bismark Hut, I built a bee house 2.4 metres square to hold just eight hives. The trees there were covered with moss and lichen and very little light reached the forest floor. That was carpeted with club mosses. Where shafts of sunlight filtered through the canopy, the drops of moisture on the hanging mosses shone like jewels; the children call it 'fairy land'.

Bee house in forest on Kilimanjaro

The next house, at 3100 metres, was more difficult to build. In dry weather one could drive up a Forest Department track through the podo forest and then through the tree-heather forest up to the grasslands. At the time of building however, we had rain and the track in the podo forest was impassable even to Land Rovers. Everything had to be off-loaded from the vehicles and carried up to the building site. Fortunately there was very little wind, otherwise the large panels of tempered hardboard would have been difficult to carry. This 2.4 metres square bee house was built in a clump of forest in a small valley in the upland grasslands. On clear days there was a magnificent view of Kibo's snow-covered peak and of the heavily eroded peak, Muwenzi. I decided to make 3100 metres the limit at this stage, because of the difficulty of access to higher altitudes, though I knew that bees did live higher up. At Peter's Hut, 3700 metres, where climbers spend the night on their way up the mountain, the thunderbox was occupied by bees.

We built the lowest bee house at 1200 metres among the coffee and banana plantations, so that we could obtain a comparison with the higher bee houses. To ensure that we had bees accustomed to each particular altitude and vegetation, we stocked the hives with bees which had occupied log hives near each house.

All this took time. There was very poor flowering that year and the bees were slow to build up. In fact I was not able to get very far into my project of determining the prospects for migratory beekeeping up and down the mountain before other events brought my personal involvement in that project to an end.

Bee varieties

But I did learn more about the different races of honeybees in East Africa. The common honeybee of the African tropics was first named *Apis adansonii* by Latreille, 1803, and is a variety of the species *Apis mellifera* Linnaeus, 1758. It is subject to considerable variation in the colour of chitin and hair, in size of body, length of wings and tongue, and in the width of the bands of hairs on the abdomen. Typically, and most commonly, the workers have yellow bands on the first three abdominal segments, a yellow scutellum and yellowish hairs. There may be four abdominal segments with yellow bands but graduations occur to all-black abdomen with yellow, brown or black scutellum. All colour forms may be produced by the same queen but, in the more mountainous areas and at higher altitudes, there is a tendency towards the darker forms and a higher proportion of dark workers. The colours of queens and drones are likewise subject to considerable variation. The length of forewing of workers collected at various places in East and Central Africa were found to be: minimum 8.1mm, maximum 8.7mm, colony average from 8.42mm to 8.51mm. These figures apply to bees collected from sea level on the coast to 3400 metres in Ethiopia. Altitude alone does not appear to have a significant influence on size in this variety.

The average width of worker cells, between parallel sides measured in each of the three directions on the face of the comb, was found to vary from 4.76mm to 4.94mm from centre to centre, average 4.80mm. Experiments at Tabora carried out before we moved to Arusha had shown that the African bees build comb most rapidly when supplied with medium weight foundation with cell bases of approximately the size they use in the wild. They were slower to build comb on the larger foundation made for European bees. The spacing of brood combs in the centre of the nest varies between 30mm and 32mm, centre to centre. The outer storage combs may be spaced up to 35mm centre to centre.

Various names have been given to this bee, according to the colour of parts of its body. But the variation which I myself observed within individual colonies, the progeny of the same queen, at Tabora and in other parts of East and Central Africa, leads me to believe that we should stick to the first name given to the race, in accordance with the rules of nomenclature.

When we came to stock the bee houses with bees from the locality of each house we found that bees obtained at 2400 metres and 3100 metres are large and black with narrow bands of hairs on the segments of the abdomen, very similar in appearance to the European *Apis mellifera mellifera*. The length of forewing varies between 8.7mm and 9.3mm, the average in different colonies being between 9.00mm and 9.09mm. The mean width of workers cells in individual colonies varies between 4.86mm and 5.22, average 5.04mm.

I found that these dark mountain bees are very hard workers and they forage at lower temperatures than the common variety, which we also obtained in the vicinity of 2400 metres. The latter appeared to be migratory, returning to lower altitudes when periods of dearth or very cold weather occur in the upper forest. The big black bees are particularly good at conserving stores, cutting down brood rearing drastically at the first sign of a dearth of nectar and remaining in their hives without migrating. In the lower forest and upper cultivated zone, between 1500 and 2400 metres, intermediate forms show ample evidence of crossing between the two varieties. I also found the large black mountain bees on Mt. Meru at between 2700 and 3100 metres. In view of their specialised habitat I named these bees *Apis mellifera monticola*.

At the high altitudes the big black bees were gentle to manage in the bee houses, but they did show the tendency to nervousness, to rush around the hive, that was said to have been a characteristic of the old European *mellifera*. They also suffered from an infestation of large white mites on their bodies. We brought some hives of *monticola* down to an apiary near my office in Arusha with the intention of raising queens from them and then taking them back up the mountain for mating. But we found one disconcerting thing; *monticola*, so gentle high up on the mountain, at Arusha was every bit as ferocious as *adansonii*.

I wish I could have done some work on the bionomics of the wing venation on *monticola* in comparison with *adansonii* and the European *mellifera* but I ran out of time. I did wonder whether *monticola* were descendants of European bees which might have been brought into East Africa in German times, before the Great War. Today, DNA tests could provide the answers.

A beekeeper at Tanga wrote to me about a very small bee he had found there and I went down to investigate. It appeared that this bee is common on the coast. I had seen its comb previously on Mafia Island. I found that the length of forewing is between 7.9mm and 8.4mm, the average in different colonies being between 8.13mm and 8.23mm. The average size of worker cells in individual colonies lies between 4.50 and 4.72mm, the mean of thirteen colonies measured being 4.62mm. The spacing of the brood combs (centre to centre) is 28 – 30mm, outside combs increasing to 32mm. This bee and its comb are significantly smaller that the common African honeybee and it lives in a region where there is almost continual flowering throughout the year.

It is quite the opposite in behaviour to *monticola* in that it does not store much in the way of reserves of food, readily abandons its hive and moves elsewhere when attacked by pests. Because its distribution is restricted to the coast, I named it *Apis mellifera litorea*.

Named type specimens of each of these three varieties were preserved at the Beekeeping Division laboratory at Arusha and I retained colour photographs of them.

Madrid Conference

In September 1961 I attended the International Beekeeping Congress in Madrid. My journey there from Arusha once again undermined any confidence I may have had in travel by air lines. British Overseas Airways ran late, so late that my plane to Madrid had departed some time before I reached Rome. After a frustrating wait at Rome airport for any possible flight onwards to Madrid, I was taken into Rome to an hotel for the night. It was a very modern hotel and the taps for a shower seemed non existent. I had to summon a night porter to show me where they were and how they worked; I felt a real hick from the high timber – and he did not speak English and my Italian was limited to what I learned in the army in Eritrea in 1941; not quite relevant in a sophisticated hotel in Rome in 1961.

As I had some time before my plane the next morning I took a stroll to see some of the nearby historic sites. I found that the traffic was less predictable than any I had previously experienced and that the only way to cross a road was to wait until a large number of Italians had gathered and then get into the middle of them. I arrived in Madrid 24 hours late and a ramshackle airway's bus had deposited me at my hotel. I found that because I did not arrive in time the previous day, my room had already been occupied. Luckily, they managed to find me another.

I gave papers on the need for *A Beekeeping Research Institute for Africa* and on *The Races of Honeybees in Africa*. But to me, and I think to most others, the most useful part of the congress was contact with people from other parts of the world with similar interests. The expatriate Anglo-Saxons seemed to gravitate together for dinner in the evening, which in Spin is rather later that the customary English feeding time. The English were conspicuous for queuing up for dinner long before the dining room doors opened. I got to know two Australians quite well, Keith Doull from South Australia and John Guilfoyle from Queensland, the former a scientist and the latter a bee appliance manufacturer. I also made friends with Gordon Townsend from Canada and Roger Morse from USA, who shared the expatriate Anglo-Saxon table. I was impressed with the Spanish wines but revolted by the traditional sea-food dish which appeared to consist of spiders and other unidentifiable nasties. An excursion to Toledo, the ancient capital of Spain, was fascinating and brought back memories of the reports on the Spanish Civil war I had read many years before. But I was really impressed by the Flamenco dancing which we saw on the last evening in Madrid. Somehow this conveyed to me the spirit of Spain, which I appreciated and understood. As I expressed it at the time, "The men were men and the women were women and proud of it". My return journey to Tanganyika was uneventful.

Overseas visitors

Back in Arusha I received a most interesting visit by three gentlemen from Angola, Dr Salagata, Eng Falgata and Virgilio de Portugal Araujo, with whom I had had correspondence on stingless bees. Virgilio had made a study of the Meliponidae in Angola and I was able to show him a nest of our *Trigona ferruginea* in a log hive in the forest above Arusha.

Another visitor I had at this time was Mr Tecwyn Jones, a Forest Entomologist working on timber borers. Apparently the visit so aroused his interest in beekeeping that eventually he was nominated as Adviser on Apiculture within the Overseas Development Administration and was elected to the Council of the International Bee Research Association, becoming its Chairman in 1983.

After ten year's experience of frame-hive beekeeping in Tanganyika I had arrived at what I considered the ideal frame hive to used with

African bees. The Utilization Division of the Forest Department determined the most suitable local timbers for hive making and worked out the mass production techniques and costings of hive manufacture. The results were published by the Forest Department in a pamphlet, *The African Dadant Hive.*

In August 1961, the Oxford University Press invited me to write a short book on beekeeping and honey production for the Oxford Tropical Handbooks series. As I had drafted such a book when on leave in 1953, and had previously offered it for publication to Longmans after they had published my text book *Beekeeping in the Tropics,* I was able to respond readily. The book was eventually published in June 1963 by OUP under the title *Beekeeping.* Some 2770 copies were sold before it went out of print in 1968. Because I left Tanganyika in 1962, I was not able myself to promote it in the tropical beekeeping regions for which it was written.

The need to move

I had not been in Arusha long when it became clear from political developments that I would have to find another job at the end of the current tour, in about another two years, unless events resulted in the termination of the services of the British officers earlier. His Excellency the Governor of Tanganyika had tried to reassure us that our services would be needed for at least another 15 years. That would take me to retiring age, but I could see that there was no future for my children there. I had enquired about a job that was being advertised in Canada, but I would have to complete my tour unless pensioned off earlier.

In August 1960 I drew up a list of advantages and disadvantages of continuing service in Tanganyika. The principal advantages were that there was plenty of scope for research, the climate was, on the whole, pleasant and the countryside was beautiful, the flora and fauna fascinating, the country was not overcrowded and we had many friends there, among all races.

Looking at the prospects for development of beekeeping, had it not been for the recent political changes in East Africa starting in 1953, I would have had every expectation of both Europeans and Africans building up bee farms during the next ten to fifteen years. But now I could see little prospect of ever getting bee farming going or of research bearing fruit in the field. Few of the European farmers were young enough to start, all were uncertain about the future. The Africans were not yet ready and their apparent increasing lack of respect for property was an inhibiting factor and would make bee farming too precarious. To the question – "Would I myself establish a bee farm in Tanganyika?" – the answer was a clear "No".

There were other disadvantages, which were confirmed when I reviewed the situation in October 1961. There was a lack of any future for European children. There was growing discrimination shown against Europeans resident in East Africa. The cost of living and of education was already crippling; it was impossible to save even when living frugally. My own professional future was uncertain and certainly limited with no hope of promotion. Doctors were already leaving, and collapsing medical services would lead to misuse of drugs and corruption; without reliable medical services and dependable drugs, European existence in the tropics would be precarious.

I was rather bothered about the non-arrival of mail, and a boycott on mail and goods from South Africa was developing. I investigated the possibility of going to the United States of America and Joan and I visited the US Consul in Dar es Salaam. Apart from being put off by the roaring air conditioner and the bitter cold of his office compared with the normal heat outside, we were appalled at the bureaucratic detail of information required on the application forms, including a report from the chief of police in every place we had lived during our adult like. There was also talk of quotas of immigrants from various countries. That, and Joan's previous experience of Americans when she was serving in the British army alongside them in Supreme Headquarters during the war, put us off the USA idea.

We began to look at the problem from the point of view – "Where can a man go, whose native tongue is English, and where there is a future for his children?" We ruled out Rhodesia: I was offered a job there, and I ruled out South Africa even though my work was well known and appreciated there. We were too uncertain about the politics of Africa, and we wanted to go where we could put down roots. We did not want to have to start again in another few years. Besides I was now over 40 years of age; I had to re-establish myself as soon as possible for the rest of my career.

The United Kingdom did not appear to have any openings. We felt that I would be too restricted there after spending all my adult life in the Army and in Africa. Joan voted Canada as too cold after our 13 years in Africa. This left Australia and New Zealand.

On the 6th December 1961, Tanganyika got Uhuru, or Independence or, as some people say, Freedom. The Union Flag was hauled down and the new Tanganyika flag raised. The new National Anthem was played. It was a nice frisky tune composed by the Police Bandmaster and immediately became known as 'Webster's Reel'.

Then things began to happen which we did not like at all. About six Europeans were expelled, not for any crimes but for imagined slights against Africans or against the new State. At dawn on the 1st January 1962 I arose from my bed with a New Year's resolution – "We go to Australia".

That very day I wrote to the Commonwealth Scientific and Industrial Research Organisation (CSIRO), the Department of Agriculture in each State, and to the University of each State, stating that I should like to immigrate to Australia within the next two or three years. I did not know that some States had more than one University. From all I received the most courteous and helpful replies. And the Department of Agriculture in Western Australia said that my enquiry was of interest and asked what salary range would interest me and requested names of referees. After further exchange of correspondence, the Department of Agriculture on 20th March 1962 stated that it was now possible to make me a firm offer and supplied details. I discussed the matter with my boss, Robert Sangster, the Chief Conservator of Forests, and he said I could go whenever it was convenient; Government would not require me to work my six months notice.

So we booked to depart on 7th May for some much needed leave and to wind up our affairs in the UK. We would then fly on from London to Perth in Western Australia by Qantas on 28th June. The children would return to their schools after Easter for the final term of the school year and join us in Perth on 31st July. While they were at Arusha for the Easter holidays we all had medical examinations and X-rays, which were required by the Australian Government.

One of the last things I did before leaving Tanganyika was to prepare a review of my work there, with suggestions as to what could be done in the future. I gave it the title of *Development of the Beekeeping Industry in Tanganyika*. While I could not complete the various projects I had started in Northern Province, I expected that my staff would carry on and bring all the work to satisfactory conclusions. At least I had laid the foundations upon which they could build.

We finished our packing and saw our boxes on the train to Tanga for shipment to Australia. We sold Joan's much loved Citroen light-15, and traded in our English Ford Zephyr for an Australian Ford Falcon which we would collect in Western Australia. When all the goodbyes were said, we were driven to the airfield in a Government Land Rover, nearly being run down on the way by a Police vehicle coming out of a side street. We boarded the East African Airways DC3: it was with a sense of relief, mixed with sadness, that we left the ground and headed for Nairobi.

So ended nearly thirteen years of service in Tanganyika, which had begun in 1949, and which had had a tremendous influence on the lives of each member of our family.

CHAPTER 9
1962: A NEW WORLD

The hired Land Rover was waiting for us at the airways terminal in London. As there were just the two of us, we would be travelling light and staying at hotels. We like to be able to brew up and have a cup of tea whenever we take a rest from driving. I found the ideal kit for the job: a little kettle that fitted on a minute stove, which used solid meths for fuel. This we could use safely in the back of the Land Rover.

Joan and I drove down to Pennant Caravans near Southampton and collected our melamine plates and mugs and other useful items from our caravan. The damp of the storage had taken the curve out of the handle of my beautiful German walking stick so I abandoned that. I arranged for the caravan to be sold and we headed north for Scotland, to Torphins and a tour of the Highlands, to relax and recuperate from the stresses of recent months, and to build up strength for whatever lay ahead in Australia.

Before us lay the prospect of spending the rest of our lives on the other side of the world. Those seven weeks of travelling quietly round England and Scotland would be a period of adjustment and reorientation. We would be saying goodbye to the land of our birth, the country which we had both helped to defend against Hitler, which both our fathers had helped to defend against the Kaiser, and which had given us our chance to make our way in our future life.

One of the first nights out we spent at an hotel at Thetford, not far from the original seat of Joan's Bardwell family. The hotel was ancient, half timbered and of mud and wattle construction. A section of wall in an upstairs passage had been opened up to show the method of construction. The floors were surprisingly uneven but the hotel was very comfortable and the food good.

We had crossed into Scotland when we spotted a convenient picnic area near the main road and pulled in to brew up. To our astonishment we were hailed by a man and his wife we had last seen a few months ago in Tanganyika.

Saying goodbye to Scotland

In Edinburgh we stopped to inspect the shops in Princess Street. After returning to the parked Land Rover, we realised we had seen no signs indicating the next section of our route to Aberdeen via Dundee. Our maps gave no clue. Hailing a passing policeman, I asked him to direct me to the road to Dundee. He looked at me is disbelief. He obviously thought I was having him on. I looked down Princess Street and every head gleamed some shade of red in the sunlight. He had taken me for a native; Joan alone stood out as a stranger with her dark Spanish hair and complexion. I explained that we were visiting the country from abroad, and he unbent enough to give us the necessary directions to set us on our way.

Up in Torphins we stayed at the Learney Arms, the one and only pub. Our goodbyes seemed rather distressing and when I visited the

University, hardly any one I knew was around.

We headed into the Highlands and stayed at an hotel in Aviemore. We climbed up to the top of Cairngorm, skirting the edge of a snow-filled corrie on the way. That was quite the most vigorous exercise we took on that leave. But the view from the top was worth it.

We made our way north to Inverness and on to Black Isle where many years before there had been a house for sale which Joan fancied as the right place to found a dynasty; it had seven bedrooms. But even in May the area seemed rather too bleak. Further on at Strathpeffer we examined a very ancient fortress where part of a wall had been subjected to such great heat that the rocks had melted and become vitrified. The cause remains a matter for speculation. Then back to Drumnadrochit and the Great Glen for a drive along the side of Loch Ness and onwards to the memorial to the Commandos near Spean Bridge. This is quite the most impressive war memorial I have ever seen. It consists merely of the figures of three commandoes, looking up across the Leanachan Forest to Ben Nevis, with the inscription from the psalm,

I will lift up mine eyes unto the hills.

We spent several days at the little hotel at Glenfinnan at the head of Loch Shiel. Here we walked in the hills, drove up the winding road to Arisaig and Mallaig, with views west across the sea to the islands of Eigg and Rhum and away to Skye to the north west. At the hotel we had most interesting company, a lawyer from Glasgow still suffering from wounds sustained in the war, a woman don from Glasgow University and an American negro woman artist, each most highly talented and very good company, once they were sure that we were not Campbells. We were now in MacDonald country.

We went back to Fort William, down Lock Linnhe, turning off to Ballachulish and up through Glen Coe where the mist and rain set the scene of the great tragedy of long ago, then across Rannoch Moor and Black Mount, down beside Loch Lomond, by-passing Glasgow, through Kilmarnock to Dumfries, where we stayed a while to catch up with Matron, who had retired there from the hospital in Torphins.

From here we made our way back rapidly south, called in on Dr Eva Crane at Gerrards Cross and left with her our collection of large scale maps for the use of other overseas visitors to Britain. We went on to Worthing, staying at Joan's father's local pub, and preparing for our final goodbyes to him and to my Uncle Lionel. Both were now alone, Mabel Bardwell had died at the beginning of our last tour in Tanganyika, and Auntie Maud was dying in a nursing home.

My little kettle and its solid meths stove had done its job; we would no longer need it, so I gave it to our maid in the hotel. We had our last drink with Joan's father in the pub and then we were in the safe hands of Qantas, flying first class in a Boeing 707 heading for Australia, with Joan having learned discussions with the steward about the merits of Australian wines. We were beginning to come to grips with our new world.

*   *   *

It was late in the evening of 28th June 1962 when we arrived over Perth and descended through the winter clouds. From the aircraft the coastal plain below us seemed to consist of lakes, large and small, gleaming in the twilight. Perth appeared waterlogged and our spirits were somewhat dampened. The plane landed and as we walked across the tarmac towards the terminal buildings, Joan and I heard what seemed a familiar voice calling, "There they are", and there was Maurice Mulcahy, last seen when we graduated together at Aberdeen in 1949. With him was the Deputy Director of Agriculture, Leo Shier and his wife.

While we were waiting for our baggage to be unloaded, Leo Shier began briefing me about the situation in the Apicultural Section of the Department of Agriculture. He must have talked for two hours. I, having flown straight out from England during the last 24 hours, was not in a position to absorb very much. But I did appreciate that there was some sort of crisis, and I would have to sort it out.

I had known that I would be taking over the Section from a man who had been in charge for some time, and I had been somewhat apprehensive as to his reaction to an incomer being brought in over him. Eventually Leo Shier finished his briefing and Maurice was able to take Joan and me off home where his wife Patricia had been preparing dinner for us, and then to our hotel in Perth where we fell into bed.

The hotel, The George, was pretty rough but the food was acceptable. Here we were introduced to the arrangement of separate ablution blocks for men and women – ensuite facilities were in the future in western Australia.

The next morning Leo Shier picked me up from the hotel and took me over the river to the Department of Agriculture, which was in new buildings in south Perth, but not before he got himself lost on the way on the new Narrows Bridge interchange. I was put on the strength immediately, completed all the administrative procedures, and was introduced around the Department. Then the weekend intervened and Joan and I were able to get a breather and take stock of the situation.

On the Monday morning Leo Shier introduced me to the Director, Dr Tom Dunne, and to the Minister, Crawford Nalder, whose offices were in the old Treasury Building in the centre of Perth. The Minister introduced me to the Beekeepers' Section of the Farmers' Union while opening their Annual Conference. This conference went on for three days and was a most valuable introduction to the problems of the beekeeping industry in Western Australia. I had to make a short address to the beekeepers and Tom Powell, the Manager of the Honey Pool, said in his reply something to the effect that I was more Australian than they were. I must have been more direct in my remarks than I reasised and my words had hit home. Tom Powell was always a good friend in the years which followed.

Perth, July 1962

I found that I had two men to assist me in the Apicultural Section. The senior was Bob Coleman, who previously had been in charge and who did the extension work, maintained the small departmental apiary, and did some apiary inspections. The other was Alan Kessell, who was mainly responsible for the apiary inspections and disease control work, and did quite a bit of extension work at the same time.

In a letter to Eva Crane early in 1963, I gave my first impressions of the industry:

"There are some 700 beekeepers in this State, owning about 50,000 hives. About 70 beekeepers are full time commercial apiarists, and they produce most of the honey, averaging over 200 pounds of honey per hive per annum. Production is now about 5,000,000 pounds of honey a year, worth a quarter of a million pounds. The price paid to the beekeepers by the Honey Pool (the producer market co-operative) this season was 7.55 pence per pound, for honey supplied to the Pool in bulk. This is less than the Africans get for their good honey in Tanganyika. It is hoped that with the formation of the Australia-wide Honey Marketing Board, the cut-throat competition between honey exporters will cease, good standards of quality will be laid down, and the beekeeping industry will be in a better position to deal with the English Honey Importers Association. I can see no development taking place in the industry until better prices are obtainable for honey. Very few of the beekeepers enjoy the standard of living of the English commercial beekeepers.

"My immediate research programme deals first and foremost with a study of the bee forage in this State with the object of producing a book on the honey flora in due course. Other problems for immediate investigation are the colour and density of WA honey to determine the causes of low standards in these respects, and the bleachability of the local beeswax to find out why difference in this quality occurs.

"I must say here that the quality of honey produced by the better beekeepers is every bit as good as any produced in England, and there is no justification for the payment of such low prices for WA honey of the better qualities. As for the Eucalyptus flavour that people in England talk about, it is a myth and a delusion, and every major species has its own characteristic flavour. I am afraid that what I have learnt about the English honey importers since I have been here, has not increased my regard for them."

I went on to say,

"Perth is a very beautiful city and the people are charming. The beekeepers are as fine a crowd of people as you could meet anywhere."

Our first personal requirement after arrival and installation in the hotel, was to get a vehicle. We obtained a Ford Falcon as had been arranged when we traded in our Ford Zephyr in Arusha. Unfortunately, although the data in the brochures showed all the technical specifications to be identical to those of the Zephyr, the Falcon did not feel the same nor perform as well as the Zephyr. However it was a set of wheels, absolutely essential for moving around Perth.

The next essential was to rent a house so that we could get out of the hotel and have somewhere for the children when they arrived at the beginning of August. It happened that one member of the Department of Agriculture was on his way to the Sudan as we arrived, so his house was available for six months. This was in Essex Street, Wembley, and we were very happy to take advantage of the offer while we looked for a house of our own.

Looking for a house was more difficult than we had imagined. We needed to decide in which area we wished to live. Driving round the suburbs, they all seemed the same except that some of the older areas close to the city had smaller blocks. Gradually we began to recognise subtle differences and that in a few areas blocks were bigger than the average quarter acre. Also on roads overlooking the river, some of the houses were more flamboyant, 'Aziz Villas' as we called them in Tanganyika. But they were side by side with much more modest houses. Eventually we decided that the Nedlands-Claremont area would suit us best. It was close to the Catholic schools to which our children would go, and to the University to which we expected they might go eventually. Also the drive from that area to my place of work, the Department of Agriculture in South Perth, was fairly straightforward.

We scanned the papers for advertisements of likely houses. We found we had to learn a new jargon: w.t. – meant window treatments, curtains or venetian blinds; w.w. – meant wall to wall carpets; u.m.r. –

under main roof, referred to laundry or loo, especially the latter; b/tile – brick and tile; con. d/r – lounge (sitting room) connected to dining room; b/r – bed rooms; b.i. cupboards – built in cupboards; s.rec – shower recess; el f. – electric light fittings; s/o – sleep out (caged-in verandah); w.i. robes – walk-in wardrobes; h.w. – hot water.

One common feature was that, if the loo was not in a separate little house down at the bottom of the garden, it was on the back verandah; you had to go outside to get to it. This was in compliance with the building regulations, which had been amended, not long before we arrived, to permit a door direct from inside the house to the loo. Not all houses were connected to the main sewerage system – this is still true today, 1992 – many houses had their own septic tanks, which required pumping out from time to time.

We inspected numerous houses, including one which was large enough for our family and had a windmill on a bore and a large water tank. Although it was going at a very reasonable price, £5,000, it would have needed a lot of money spending on it to bring it up to scratch. Also it was on the busy Stirling Highway in Claremont. Soon afterwards it was bought as a training centre for Air Force Cadets and an aeroplane was mounted in the front garden. Recently has been sold for $2,000,000 for development into a shopping mall!

Then we spotted a small quiet advertisement:

NEDLANDS: Lovely spacious 23 square b/tile in high pos. Close all amenities. Grounds are fully reticulated from own well - - - . Price £9,730 Phone owner at - - - . We phoned. The owner was Ernest Lennon, Auctioneer and Real Estate Agent. He was arranging to move into a bigger house in Mosman Park. We inspected the house and grounds and fell in love with it, putting down a deposit immediately.

Then came the problem of paying for it. The cost was covered by my compensation for loss of office in Tanganyika, but that was being paid out in five annual installments. The bank with whom I had opened an account in Perth sent out a valuer who said it was worth only £7,000. Ernie Lennon introduced me to his bank manager. I transferred my account. I got what was needed to cover the cost, at what I considered a reasonable eight per cent interest on the debit balance in my account. I discovered that Ernie, as an Australian ex-serviceman, was able to borrow the balance he needed for his new house through the Veterans' Affairs Department at a rather lower rate of interest. Such facilities were not available to British ex-servicemen. Two weeks after our four children had arrived from their schools in East Africa we moved into 36 Vincent Street, Nedlands, and have been there ever since.

Now, with a car and a house we were getting organised. Next, to find places in schools for the children. Rosemary was no problem at all; she merely transferred from Loreto Msongari in Nairobi to Loreto Osborne in Claremont. In a sea of blue uniforms, wearing her Msongari blue hat and scarlet blazer, she was quite conspicuous. The nearby St. Louis boys' school could take David but not Christopher. Not wishing to separate them, we booked them both into Trinity, which was a rather longer bus journey. Antony started at the nearby St. Teresa primary school before going on later to Trinity.

One little hiccup we encountered was that while the school year in East Africa ended in July, in Australia it went through to the end of the calendar year. This meant that the children found themselves plunged into the middle of the curriculum for the year and, added to that, the Australian curriculum was different from the British one. Somehow, they survived, more or less.

At work my most urgent task was to prepare a paper on the beekeeping industry in WA and on how the Department could assist it, in the meantime visiting the leading beekeepers, touring the main beekeeping areas and familiarizing myself with the local flora. I saw some of this on the road from Perth to Yanchep, where the Department of Agriculture's bees were kept. My first impression was that I was on a different planet. The plants were so vastly different from those of Europe and Africa. But then I began to see a few things that had a connection with some African plants.

One job the beekeepers were keen for me to do was to examine the honey producing potential of the vegetation in the Goldfields, some 400 miles to the east of Perth.

My chance to make a start on this occurred very early when the National Parks Board asked us to move the Department's apiary at Yanchep, where it was in the way of the development of facilities for the golf course. I went out with Coleman to see the nature of the problem and was horrified to find an apicultural slum. I ordered the immediate removal of the hives to the Eastern Goldfields and a thorough cleanup of the apiary site. I think the staff were delighted. They had never dared to suggest such a radical move. The National Parks Board was delighted also. I learned that Clee Jenkins, the Board's President and an entomologist, did not like bees; whether this was because of their tendency to sting or for ecological reasons I was never quite sure.

CHAPTER 10
LEGISLATION

It was only when I read the files that I began to realise what Leo Shier had been talking about during those two hours at Perth Airport. The Department of Agriculture was in a mess with the old *Bees Act, 1930-1957*. Amendments made to the Act over the years confused rather than clarified the meaning of the Act and now the Department was being threatened with legal action by one very astute commercial beekeeper. It became apparent to me that the immediate trouble had arisen mainly through a clash of personalities between the beekeeper and the inspector.

The Apicultural Section had been transferred before my arrival from the Horticultural Division to the Animal Division. I was therefore working directly under the Chief Veterinary Surgeon, Claud Toop, who was also the Chief Quarantine Officer for the State of Western Australia under Commonwealth Animal Health legislation. Claud Toop was most helpful and supportive of my efforts to sort out the legal problems.

My first trip into the Jarrah forest was to inspect the apiaries which were the subject of the argument and in which American Foul Brood disease had been detected. They were in the Collie area near the Wellington Dam. I met the owner of the hives in his apiaries and heard his side of the story. I found that I was able to pacify him with regard to his personal conflict with the inspector but I also pointed out to him that he had broken the quarantine regulations and laid himself open to prosecution, which was something the inspector had failed to get across to him. That concluded what had been a very tiresome affair to the Department and I was able to get down to revising the whole Act and its Regulations, the weaknesses of which were now very apparent.

My previous experience with the Scottish Beekeepers' brood diseases of bees insurance scheme, the preparation of the Tanganyika *Produce Export (Beeswax) Rules, 1957*, and the drafting of the *Prohibited Import (Bees, Beeswax Honey and Beekeeping Appliances) Order, 1962*, under the East African Customs Management Act, stood me in good stead in drawing up the new act and its regulations. Den Sander of the Crown Law Department was most helpful in doing the final drafting for Parliamentary approval without altering the sense of any clause.

The purpose of the *Beekeepers Act, 1963* was to provide the beekeeping industry and the authorities responsible for the administration of the act, with an easier and more effective means of controlling and eradicating bee diseases and pests, of maintaining the orderly conduct of the industry, and for the improvement of the products of beekeeping.

The new act included many of the provisions of the act it was replacing, but phrased them more clearly, as well as covering aspects which had not been considered previously.

Under the new act, everyone who owned hives of bees, or who managed or had charge of bees, even if they were not their own property, or had charge of hives which had had bees in then in the past, was obliged to register as a beekeeper at the Department of Agriculture and renew the registration each year. When he registered or renewed his registration, the beekeeper was required to pay his contribution to the Bee Industry Compensation Fund under the *Bee Industry Compensation Act*. This is a form of insurance against the loss of bees and equipment arising from disease eradication measures. Whenever a beekeeper leased, sold, exchanged or otherwise disposed of hives of bees he was required to notify the Department of Agriculture.

In order that hives and apiaries could be identified, every hive owned by a beekeeper was to be branded with a registered brand, consisting of one or more letters and one or more numerals. Each brood box and honey super was to be branded and the brand could be used, if the beekeeper so desired, on other hive parts such as bottom boards, lids and frame top bars.

Each beekeeper, when he registered as a beekeeper for the first time, was required to apply for a brand, which was issued and registered by the Department of Agriculture. When a beekeeper acquired hives which were already branded, he was required to place his own brand below that of the previous owner and score a line through the old brand, but without obliterating it. This could provide valuable information when tracing back to the source of a disease outbreak.

The only type of hive which was allowed to be used was one having combs built in movable frames, which can be removed easily for inspection. The use of boxes, barrels or other containers, in which the combs cannot be removed easily for inspection, was totally prohibited.

In the absence of a natural source of water near the apiary, such as a pond or stream, beekeepers were now obliged to provide near every apiary site a good supply of water readily accessible to the bees. All too often, complaints had been received from farmers about bees collecting waster from troughs and preventing stock from drinking, and from others who found that swimming pools or garden hoses had been invaded by bees. The provision of water is essential for the well-being of the bees in most parts of the State.

The act now required beekeepers to keep their bees in such a manner that they would not be a nuisance to neighbours or to passers-by on public thoroughfares. This included refraining from putting apiaries close to roads where the bees could pose a danger to passing traffic or to stock.

Also, when moving bees, the new act required beekeepers to make sure that they would not be a nuisance to other people and, when carrying hives of bees with open entrances, to avoid such practices as stopping in inhabited areas or at petrol stations, or moving on public roads in daylight.

In cool weather and using ventilated migratory lids, hives of bees can be moved with closed entrances. But many apiarists in Western Australia, using mechanical loading devices, found it convenient to move the bees with open entrances. It is then necessary to load in the late afternoon and for the truck to wait in the apiary until all the bees have returned before moving off at dusk. Care had to be taken not to halt near lights, and the engine is kept running to stop the bees from crawling all over the truck during halts. The bees are usually unloaded at dawn on the new site. If the journey takes more than one night, the truck-load of bees is parked in the shade of trees during the day and the bees provided with water.

One of the most important new provisions of the act was that a beekeeper must report to an inspector the occurrence of disease in his apiary immediately he finds it. Every registered beekeeper was supplied with a copy of a bulletin describing and illustrating the symptoms of foul brood diseases, so that there could be no excuse for not knowing what foul brood looks like.

The new act obliged beekeepers to take steps to eradicate diseases and pests. Details of the procedures were contained in the regulations and in the bulletin on *Foul Brood Diseases*. Provision was made for the destruction of infected articles and for the isolation of infected apiaries in quarantine.

In addition, the act required the beekeeper to take the necessary steps to prevent wax moths from breeding in any stored or discarded combs of wax and, if the wax moths should occur, to destroy them and their eggs, larvae and pupae.

Unfortunately, one conscientious beekeeper with a large stock of stored supers, sought the advice of a chemical firm without reference to the Department. He was recommended to use a product which contained Dichlorvos. It protected the combs from wax moth alright. But nobody had told the salesman or the beekeeper that wax absorbs Dichlorvos. When he came to put the supers on his hives, the Dichlorvos killed his bees.

The importation of bees, combs, used hives, honey and used appliances was controlled under the *Beekeeper Act*. But the standards of quality of honey offered for export were laid down under the *Export (General) Regulations*, which applied to the whole of Australia and were administered by the commonwealth Department of Primary Industry.

Other legislation of importance to beekeepers included the *Honey Industry Act, 1962*. This provided for the establishment of the Australian Honey Board to promote and control the export of honey, to promote trade and commerce in honey , and to encourage the production and use of honey. The act provided for the licensing of exporters and the laying down of conditions under which honey might be exported. The Board was financed by the *Honey Levy Collection Act, 1962,* and the Honey Levy acts; these provided for the collection of a levy on honey, initially of ½ d. per pound, paid by the honey packers, dealers, agents, producer-packers and manufacturers using honey.

Many apiaries were on private land by arrangement between beekeeper and owner. Apiary sites in Forest Reserves, Crown Land and other reserves and parks were controlled by the Forest Department, to whom application had to be made. An Apiary Site Permit was issued subject to regulations under the Forests Act. The main provisions of these regulations were that the applicant must have at least 25 hives of bees and might not hold more than four apiary sites for every 50 hives in his possession. No person could be granted a permit for a site within three kilometres of a site granted to another person. A deposit was paid on application for the site and there was an annual rental. The Forest Department has since been taken over by the Department of Conservation and Land Management (CALM).

Each District Forest Office maintained a map showing the apiary sites in the area under its control, but the actual issue of permits was done by Headquarters in Perth where the master maps were maintained. The Forest Officers were most helpful to the beekeepers and were very knowledgeable about the distribution of honey-producing trees in their areas and about the flowering conditions. Areas in which the presence of diseased wild bees was suspected were also noted on the maps.

The new Beekeepers Act put the onus for detecting and reporting the presence of disease on the beekeeper himself. It also required him to take the necessary steps to eradicate disease and pests. On the whole, the system worked well, but as always, there were a few beekeepers who failed to comply with the spirit of the act. They omitted to report the disease and they used antibiotics in an attempt to cure the outbreak. The antibiotics enable the bees to clear the combs of diseased brood and prevented the development of the bacteria but did not remove or destroy the spores which remained in the combs and elsewhere in the hive. The continued use of this hive material led to the spores being spread throughout the apiary and to a more widespread outbreak of the disease when the use of antibiotics ceased. Such beekeepers were a continuing danger to the industry.

In some areas the existence of infected bee colonies in hollow trees was a problem. If such a situation was suspected, we put Departmental hives in the area to see if they picked up the infection. If present in a tree, the disease was eliminated only by a severe bush fire or by felling and burning the hollow tree or trees.

The WA Honey Pool had been established in legislation some years earlier as a Producers' Marketing Co-operative. Under the terms of its constitution the Honey Pool had no option but to accept any honey from a beekeeper, whatever the quality. When I tried to improve the quality of honey being produced, I found that I was confronted by the same difficulty that I had had with beeswax when I arrived in Tanganyika. If you can sell a bad product, then some will produce it. But the beekeepers who controlled the Honey Pool had no desire for change.

CHAPTER 11

EXTENSION WORK

Once the new Beekeepers Act was in force, the regulatory aspects of the work of the Apicultural Section seemed to be straight forward. To ensure that every beekeeper could recognise foul brood disease and report it, should it occur in his hives, I wrote a new illustrated bulletin on the subject and distributed it to every registered beekeeper.

I visited beekeepers in their apiaries when they were extracting honey; I was particularly interested in the equipment and techniques they used in their mobile extracting caravans. These plants for extracting honey and rendering beeswax in a confined space were most ingenious, but the processes used in some were highly undesirable; they damaged the quality of both honey and beeswax.

The first thing which struck me was the haste with which some beekeepers extracted their honey. As soon as the bees filled the supers with nectar, it was extracted without waiting for it to be converted into honey. Also the majority of the combs in the supers were darkened by the rearing of many generations of brood. The reason for this was the lack of sufficient supers. The hives were merely two-deckers, each consisting of two 10-frame or two 8-frame Langstroth boxes.

The better beekeepers added more supers as required and give their bees a chance to ripen the honey before they extracted it. They were producing a very good product.

The centrifugal honey extractors most favoured in the mobile plants were the semi-radials with reversible baskets. They are compact and efficient. Power to run the plant was usually provided by a small electric generator, placed at a distance from the van; some vans merely had a petrol engine mounted on the tow bar, with a pulley drive to the extractor and the honey pump. Some extractors were fitted with a steam-heated pipe round the inside of the barrel of the machine. This heated the honey as it flowed down inside the extractor. The owners considered this necessary to enable the centrifugal honey pump to work effectively. Gear pumps for honey had not reached Western Australia at that time.

The most common method of removing the wax cappings from the combs was by means of the hand-held uncapping knife. There were some mounted vibrating knives and uncapping machines were beginning to appear in central extracting plants. With the knife running along the sides of the top and bottom bars of Langstroth standard frames, between 10 and 25 per cent of the honey was taken off with the wax and went into the cappings reducer.

Most of the mobile extracting plants were equipped with steam generators which heated the uncapping knife, the cappings reducer and the heating coil round the extractor. A few beekeepers had installed hot water circulating systems to do the same work.

I saw the steam-heated cappings reducers or cappings melters as being the worst defect in the extracting plants. By melting wax in contact with honey, the steam-heated reducers not only coloured the honey with dark matter from the brood combs but the high temperatures also produced browning (5-hydromethylfurfuraldehyde or HMF) and distinctive flavours from both sources.

Up to 25 per cent of the honey crop of some beekeepers was subjected to this treatment and was then mixed in with the less damaged honey from the extractor.

But the problems did not end there. The honey crop was run into 44-gallon (200- litre) galvanised drums which often stood out in the sun until taken away to the packer; more heating of the honey and more HMF production.

Then there was the problem of the beeswax. Why was so much of the wax produced by commercial beekeepers a dark green-grey colour and resistant to bleaching? It was nothing to do with the flowers which produced the nectar or the pollen which gives beeswax its natural colour; that was demonstrated by Rob Smith, a leading beekeeper who consistently produced beautiful cakes of yellow beeswax. The dark stained beeswax reminded me of the beeswax produced in Northern Rhodesia, which had been melted by the beekeepers in iron cooking pots and by the traders in iron drums.

Clearly I was faced with a mighty task in trying to change the ways of some of the beekeepers. And it was back to the same problem that I found in Tanganyika. If the producer could sell rubbish, why should he bother to change his techniques, especially if he was paid very little more for the better product?

*   *   *

After eighteen months of working in Western Australia, I was not satisfied that my ideas for improving the quality of honey and beeswax were being passed on to the commercial beekeepers. My earlier concerns about the attitudes of the previous head of the Section were beginning to be realised. The opportunities for me to get out of the office and talk with the beekeepers were insufficient for my ideas to have any real effect, particularly with the more conservative beekeepers. So I decided to produce a quarterly journal, *Apiculture in Western Australia*, to be distributed free to all beekeepers. As I said on page one of the first issue – "The standard aimed at in content and in production is the highest and most up to date that we can achieve, in order that this journal may make a real contribution to the advancement of apiculture in Western Australia." While I realised this would have no effect on those who would not bother to read, I hoped I could get through to the leading beekeepers and to the innovators, and through them, in time, my ideas might filter through to those who most needed to improve.

*   *   *

In my little journal *Apiculture* I launched into the subject of honey quality. Following a complaint from the Manager of the Honey Pool I struck out with the following: - "HONEY? Did the person who had filled a truck load of drums with liquid during the first ten days of the Marri flow, really believe that he had extracted honey?" Then I came across an earlier report about a consignment of tins of fermenting unripe Karri honey which blew up on Manjimup railway station.

I warmed to my subject in a series of editorials. In the extracts that follow against the year of publication I have done some necessary editing, including converting measurements into metric units from the Imperial system in use at the time.

1964   HONEY QUALITY

To improve the quality of honey reaching the consumer at home and abroad is the biggest problem facing our industry at the present time. Prospects for expansion of the industry, the ability to sell in years of abundance and the assurance of a fair price for our product, all depend upon the attainment and maintenance of high standards of quality.

When the bees are given the chance, they make excellent honey from most of the plants in our native bush and our farms. But all too often, they are not given the time to complete the process of turning nectar into honey, or are they are not provided with the conditions under which they can do their best and most productive work. Flavour, aroma and moisture content are vital factors in honey quality. Colour is often an indication of flavour or of damage caused by processing or storage. But colour alone should never be allowed to be of more importance than flavour, aroma and moisture content.

The first causes of the production of low quality honey lie in faulty beekeeping techniques. If hives are robbed before the nectar has been converted into honey, then the product will have a high moisture content. It will be thin and liable to ferment, and will lack the delicious flavour and aroma which honey should possess. If honey is extracted from dark combs which have been used continually for breeding it will have a darker colour and a pronounced brood-comb flavour, quite unlike the true floral flavour of honey from white combs.

It is remarkable that in an industry which in many ways has reached a high level of technical development, there are still beekeepers who obtain they honey crop by robbing hives consisting only of a two-box brood nest. Such beekeepers may claim that they produce as much honey as others who used honey supers and for less capital cost, but that is very doubtful. Robbing such hives involves far more labour to the beekeeper and more disturbance to the bees than the harvesting of supers full of ripe honey. In addition, there is a marked difference in the quality of the honey produced by the two systems. The remedy is to provide enough honey supers to each hive for the storage and ripening of the honey crop.

1964   TO ROB OR TO HARVEST

The term 'robbing hives' is used by some to indicate their collection of the honey crop. 'Robbing hives' conveys the idea of the most primitive way of collecting honey. It is associated with collecting from nests in hollow trees or from the simple log hives and one imagines an end product consisting of a mixture of wax, bees, brood, pollen and honey. It also conveys the idea of depriving the bees of the food they need for their own sustenance and taking anything that is in the hives, including nectar and unripe honey.

Beekeepers using modern hives are expected to manage them in such a way as to enable the bees to produce a surplus of honey. This surplus is the amount of honey stored in excess of the amount each bee colony requires for its own use. When it is ripe, this surplus honey provides the beekeeper with his crop. Many beekeepers take great care to manage their hives so that they can harvest the crop without robbing the bees. Also they are very careful to take only ripe honey.

We know that there are still some who do rob hives. They may keep their bees in frame hives, they may use machines to help them rob and they may use trucks to move their bees from one honeyflow to the next. But in other respects their methods are very little in advance of those of the users of the most primitive hives

It is not only the producers of poor quality honey who lose by getting lower prices for the product. It is the whole industry which is affected. To try to improve the poor quality honey to make it acceptable on the market, the packer who has accepted it has to blend it with the finest quality honey. This results in a product which is not representative of the high quality of honey which can be produced and is being produced by the good beekeepers.

An essential step which must be taken is to bury the practice of "robbing hives" in the past where it belongs and for all to move forward as true beekeepers to harvest crops of the best quality honey.

1966 QUALITY STANDARDS FOR HONEY

Some beekeepers produce honey which is unripe, dirty or damaged. This low quality honey is being accepted by packers. Sometimes a small deduction is made from the price, but the penalty is insufficient to encourage the producer to take care in the extraction and preparation of his honey, or to heed advice given to him on this subject.

The poor honey is blended with good honey in an attempt to produce a marketable product, to the detriment of the good honey. This is most unfair to the conscientious apiarist who takes care to produce a first class product. It is also unfair to the public, many of whom seldom or never taste first class honey, but are quite capable of recognising it when it does come their way.

In the absence of defined quality standards and of an incentive to adhere to them, no improvements can be expected in the techniques of the industry or in the quality of honey offered to the public. Further, unless there is an improvement in the quality of the honey put on the market, efforts to increase the local consumption will be frustrated.

The only effective incentive for improving the quality of honey would be that provided by the establishment of simple quality standards, which could be enforced readily. Such standards should be simple, and be such that a consignment can be judged in a few moments without recourse to prolonged and expensive laboratory analysis. Standards are needed particularly in respect of the water content of honey, its cleanliness, freedom from damage including overheating, metallic staining, fermentation and objectionable taints, and freedom from adulteration.

*   *   *

After blasting those beekeepers whose methods I deplored, I turned to constructive suggestions on the theory and practice of using supers for the bees to store honey in the hive.

1964   HARVESTING

A super can be taken off the hive as soon as the honey in it has been ripened. Once the bees have completed the task of ripening the honey, they normally seal the cells over with wax caps. When the main mass of the honey in the super has been sealed over in this manner, it is safe to take it off. Occasionally the bees do not seal the honey, even though it is ripe. This may happen if an intense honey flow stops suddenly. To check whether nectar is still being brought in, shake one or two of the outside combs. If drops of nectar fall out, leave the super on until the next visit.

It does no harm to leave the honey on the hive after the combs have been sealed; a certain amount of ripening continues to take place even after sealing. Some beekeepers leave the supers on until after the end of the honey flow and then take off all the supers. If the flow is a very long and heavy one, it may be necessary to take off supers as soon as they are filled with ripe and sealed honey, then to extract the honey and return the supers to the hives ready to receive the rest of the crop.

It the hives are going to remain on the apiary site for some time after the end of the honey flow, the removal of the supers can be left until 10 days or more after the honey flow has ceased. If the hives are interfered with immediately after the end of the flow, the bees are liable to start robbing each others' hives. But after 10 days or so, the bees in the apiary have usually settled down sufficiently to enable the supers to be removed without the bees starting to rob, if the proper precautions are taken.

However, it may be desirable to remove the supers full of ripe honey before the flow ceases, as the bees are good-tempered and less inclined to rob during the flow.

If the hives are to be moved to another apiary as soon as the flow tails off, all supers containing honey should be removed and only empty or nearly empty supers left on the hives. Any supers containing nectar can be returned to the hives on the new site.

Hives should never be transported with a lot of honey or nectar in the supers. Apart from the increased difficulty in loading and unloading the hives, there is a great danger from the honey running down from the supers on to the bees and brood, drowning them and causing panic with total loss of the colony.

* * *

Methods of removing the honey were described including clearing supers with escape boards; emphasis was given to the use of queen excluders to keep brood out of the supers. The dangers of using chemical repellents, carbolic (phenol) and others were discussed.

Precautions against robbing were emphasised: Do not leave any supers about uncovered, even for a few minutes, otherwise bees will find them, and in no time you may have the apiary in an uproar with colonies attacking each other and stinging anyone in sight. After supers have been taken off the hives, stand them on boards and cover each pile with a board to keep the bees out. Always stack supers neatly one on top of the other so that there are no cracks or gaps through which bees can enter.

Before leaving the apiary, make sure that lids have been replaced on the hives properly so that bees intent on robbing cannot get into them. Also see that hive entrances are of such a size that the bees can guard them against intruders.

1966   THE RIGHT TOOLS FOR THE JOB

The difference between a hard day's work and a good day's work is good management. And the first step towards good management is to have the most suitable tools for the job.

The most important tools in honey production, both for the amateur as well as at commercial level, are the hives in the apiaries and the honey extraction equipment.

Commercial beekeeping in Western Australia has been said to be the hardest work there is, even harder than cane cutting.

Commercial beekeeping would not be such hard work if more suitable equipment were used. Hives which may not be too irksome to use when there are only a few in the back garden, can be an intolerable burden to manage when counted in hundreds. Honey supers which are too heavy to carry when full, which feel as though they are filled with lead when stuck down with propolis and burr comb to the box below, tend to encourage beekeepers to pick the honey combs out of the hive one by one. The use of one type of frame for both brood and for honey storage makes it necessary to space each comb separately with the fingers to obtain the wider spacing which the bees use for honey storage, and the narrowness of the frame results in one-fifth of the honey being cut off with the uncapping knife and passing through the cappings reducer.

These practices have no place in modern commercial beekeeping. They are one hundred years behind the times. The shallow honey supers, 168 mm deep, now referred to locally as Manley supers, are designed for the job. They are of the best depth for clearing the bees efficiently and are of manageable weight. Many beekeepers have recognised that the Langstroth full-depth 10-frame box is too heavy as a honey super. When full they may weigh an average of 36 kg and occasionally 41 kg. The Manley supers average 25 kg gross, a much more reasonable weight for an item which is to be lifted many times during the day.

The supers should be fitted with Manley frames having the correct self spacing and width of top and bottom bars designed for efficient uncapping. Used over queen excluders, as all honey supers should be, they make what is now a hard day's work into a good day's work.

These supers are discussed in more detail in chapter 16, Hives.

* * *

From the first issue of *Apiculture* I started a series of articles under the title 'Beekeeping Techniques'. These covered such basic matters as doubling, reversing and reducing brood nests using the Langstroth equipment currently in the hands of all beekeepers. I discussed the use of queen excluders, methods of adding supers, harvesting, including the use of escape boards and chemical repellents, and honey extracting. Later the articles covered producing new queens, selection of breeding stock and details of the techniques of raising queens, mating, and the method of introducing queens to colonies.

In the second year, articles dealt with 'The Hive' in all its aspects, from types of hives, choice of hive, bee space, comb spacing, the various types of frames, comb foundation, and the construction and dimensions of hive bodies. The various types of bottom boards, hive lids and queen excluders were considered in detail. Preservatives for hives and the assembly of hive bodies and frames were described, and later the Dadant depth (Manley) honey super and the production of comb honey in sections were the subject of special feature articles.

With these two series of articles I hoped that I had been able to give readers a good coverage of basic beekeeping methods and appropriate hive equipment, and had given them ideas which they could use in improving their own methods.

CHAPTER 12
EXTRACTING PLANT

The honey extracting plants, especially the cappings reducers used in them, were my second major target in striving to improve the quality of honey. The use of steam for heating honey in the extractor itself and for operating the capping reducer was at the root of the problem.

I drew attention to all the defects I had observed in honey.

1. Damage caused by capping reducers:

    Colour darkened by contact with liquid wax.

    Colour darkened by excessive heating.

    Flavour tainted by dark liquid wax.

    Excessive proportion of honey (up to 25%) with cappings.

    Extracted honey contaminated by reducer honey.
    Contamination from old combs in reducer.
2. Damage in extractors:

    Colour darkened by excessive heating.
    Contamination from bare iron.
    Contamination from decaying brood.
3. Other damage in extracting plants:

    Contamination through lack of straining.
    Metallic contamination from iron fittings.
    Metallic contamination from iron in pump.
    Contamination from unclean equipment.
    Absorption of water from leaking equipment.

I was not alone in my opinion of extracting procedures. When Dr J.E. Eckert (former Professor of Apiculture at the University of

California) visited Australia in 1958, he saw that the main technical problems of the industry lay in the management of hives and in honey extracting. He noted that because no queen excluders were being used, much time was wasted in sorting combs in the apiary. Also insufficient supers were being used, resulting in crowded hives and loss of honey. He suggested that hives should have four or five supers and that provision of them would lead to greater yields of honey.

In honey extraction Dr Eckert commented on the damage being done in steam-heated cappings reducers and damage to extracted honey which had been heated by steam. He pointed out that portable extracting plants limited the size of commercial beekeeping operation to 450 to 650 hives. In 1960 Dr E.J Dyce (former Professor of Apiculture at Cornell University, USA) also commented on the damage done to the colour and flavour of our honey by cappings reducers.

Dr Eva Crane, in her long report on her visit to Australia in 1967 listed some possible causes of reduction in the quality of honey.

1. Bee Management:
   (a) Use of brood combs for honey.
   (b) Contamination from repellents during removal of honey from the hive.
   (c) Removal of unripe honey for extracting.
2. Uncapping:
   (a) Honey heated and in contact with molten wax in the cappings melter.
   (b) High proportion of honey with the cappings, and then mixed back with uncontaminated honey.
3. Extracting:
   (a) Excessive local heating at bottom of extractor.
   (b) Contamination with metals from scratched or damage extractor or fittings.
4. Moving:
   (a) Excessive heating to liquefy the honey so that it will run

through pipes.

    (b) Contamination with metals, as 3.

5. Storing:

    (a) Excessive local heating near the surface of drums stored out of doors at high temperatures.

    (b) Contact with beeswax particles melted locally at hot points of drums.

    (c) Over-long and/or excessive heating in hot-room before filtering.

    (d) Contamination with metals.

In the face of all these criticisms of our extracting methods, I tried to pinpoint what was happening to the colour of our honey and I produced the following report:

1968   DETERIORATION OF THE COLOUR OF HONEY

Beekeepers and honey packers have experienced the phenomenon of honey darkening in colour during extracting, processing and storage.

This change in the colour of honey after being removed from the bees results from one or more of four causes:

1. Absorption of soluble stains from wax.

The first cause of the darkening of colour is the use of dark combs for honey storage in the hive. The old dark combs contain a dark brown colouring matter which is readily soluble in water or honey. It becomes apparent when dark combs are soaked in clean water, and even more apparent when the wax is melted in water.

The effect on honey can be observed by comparing the honey extracted from white combs with that extracted from old dark combs which were filled during a honey flow from a single source. The honey from the dark combs is substantially darker than that from the white combs and has a tainted flavour.

A more dramatic darkening occurs in the cappings melter. This darkening takes place even if the honey has not been heated enough to cause damage through heat alone. The melting of the dark cappings in contact with the honey releases the soluble stains, which are absorbed by the honey just as they would be absorbed by water.

Due to the narrowness of the top and bottom bars of the frames commonly used for honey storage and extraction, up to 25 per cent of the honey crop goes through the cappings reducer with the melting cappings. When dark combs are used, the magnitude of the damage to the honey is appalling.

This cause of darkening of honey and tainting of its flavour is avoided by beekeepers who carry out the following practices:

    (a) Use only white comb for the storage and extraction of honey.

    (b) Use honey extracting frames with wide top and bottom bars so that only the absolute minimum of honey goes with the cappings into the cappings reducer.

    (c) Permit the honey to drain from the cappings before the wax is melted.

2.   Formation of dark stains by chemical reaction between honey and metal.

Chemical reaction between honey, which is acid, and metals, particularly iron or ordinary steel, can take place during extracting, processing or storage. The damage caused by this chemical reaction is greatly accelerated when temperatures are high.

Very little damage of this nature takes place in honey extracting plants nowadays, because most beekeepers are careful to ensure that their appliances are either made of corrosion resistant materials, or are suitably coated. But it does still occur in some plants.

The equipment used by honey processors and packers is likewise made of corrosion resistant materials, though some of the pipes through which honey is passed when hot can cause damage of this type if they are of galvanised iron and if the threads and cut ends of the pipes have not been suitably coated.

The main cause of the chemical-reaction-with-metal type of damage is the use of storage drums with imperfect linings, and the prolonged heating of honey in such drums, as may be necessitated when the honey has granulated. Whole batches of honey packed for retail sale and on the shelves of shops have been seen with this metallic stain.

Damage to honey caused by chemical reaction between the honey and metals can be avoided by ensuring that the following precautions are observed:

(a) Use equipment for handling honey which is made of corrosion resistant substances.

(b) Ensure that any iron or ordinary steel with which honey may come into contact is coated with an acid resistant material.

(c) Ensure that all pipe connections are suitably protected internally.

(d) Ensure that drums used for holding honey have undamaged linings, and any that are imperfect have been retreated.

3. Darkening through heating.

The third cause of honey becoming darkened in colour is the application of heat during extracting or processing. It is not just the heat alone that causes the damage, it is the time during which the honey is held at a specific temperature that governs the extent of damage.

This subject has recently been fully investigated by the author and the effect of different temperatures ranging from 43°C to 80°C on a variety of honey types has been studied. The results give the relationship between time, temperature and colour deterioration.

The deterioration was measured in the Pfund Honey Colour Grader which is graduated on an arbitrary scale in millimetres. If a difference of 1mm on the Pfund scale could be seen, this was taken as being measurable damage. The time taken to produce this measurable damage at the various temperatures is given, together with the time taken to produce 5mm and 10mm damage.

Time in hours to darken honey at various temperatures.

Temperature		HOURS to increase Pfund scale reading		
°F	°C	by 1mm	by 5mm	by 10mm
110	43	40 to 48	90 to 120	–
115	46	9 to 18	75 to 92	–
122	50	4 to 7	44 to 72	120 to 144
140	60	2.5 to 4	11 to 22	33 to 50
158	70	0.5 to 1.5	3 to 5.5	6 to 11
176	80	0.1	0.8 to 1.5	4 to 5

Note: Dryandra honey was found to darken at twice the speed.

An interesting point to arise from this research was that the time and temperature required to produce 10mm of darkening on the Pfund scale was approximately the same as that found by J.W. White et al (1963) to produce 3mg of 5-hydroxymethylfurfuraldehyde (HMF) per 100g of honey.

The table shows that when the honey is at 43°C it takes between 32 and 48 hours to produce any measurable darkening of the honey, while at 80°C the damage appears after 6 minutes.

In the honey extracting plant, damage to colour through heating can occur in the extractor, where a steam or hot water coil is fitted internally and forms a ledge which can hold up honey or wax particles. Where the coils are fitted externally, or are arranged so that all the honey flows past the coils quickly, then the time of contact with the heat is so small that no damage is done.

The other appliance which causes deterioration of colour through heating is the cappings melter. In the best designs, the honey flows away from the heat rapidly, and most of the damage is due to contact with the melted cappings and absorption of stains and taints.

In processing plants, the main cause of colour damage is the heating of honey in drums to liquefy it. Further damage may occur where a large bulk of warm honey cools slowly.

Damage to honey by heating can be avoided by the following measures:

(a) Ensure that heating coils fitted to the extractor are above the level of any honey standing in the tank and that they are on the outside of the body of the machine, or if inside, are so fitted that there is a smooth sloping surface above and below them which does not hold any honey or particles of wax.

(b) Either drain the honey from the cappings without the use of heat, or use an efficient cappings reducer which does not retain any honey in contact with heated surfaces, nor retain any bulk of heated honey and which permits the honey to flow away from the reducer rapidly.

(c) Heat honey in drums only to between 38°C and 43°C to render the whole into a readily flowing mush which is decanted through a coarse strainer. The crystals of dextrose still remaining in the honey can be dissolved by passing the honey rapidly through a heat exchanger before filtering and then cooling the honey immediately after filtering by passing it through another heat exchanger cooled by water; the whole process after decanting being on a continuous flow system.

4. Deterioration in storage.

Darkening in colour, increase of HMF and destruction of enzymes, is rapid in storage when the temperature is above 30°C. Damage to honey stored in galvanised drums in the sun will be greater owing to the excessive heat absorbed by honey in contact with the metal.

For the beekeeper, shading the drums from the sun will help, and is quite easily done. The packer, with large numbers of drums to store, perhaps for several months, not only would need a shaded storage space, but also would need to have an air conditioned store for honey which is held for long periods.

The rate of darkening in storage is governed by temperature and time as it is in processing.

The following table shows the rate of darkening per month at various temperatures.

Rate of darkening of honey in storage

Storage temperature		Rate of darkening per month
°F	°C	(increase in mm Pfund scale)
50	10	0.024
60	15.6	0.080 to 0.125
70	21.1	0.270 to 0.700
80	26.7	0.900 to 4.000
90	32.2	3.000 to 7.700
100	37.8	10.000 to 14.000

At 15.6°C the honey takes between eight to twelve months to darken 1mm on the Pfund scale. At 21.1°C this amount of deterioration takes between six weeks and four months; at 26.7°C between one and four weeks. Six months of storage at 26.7°C average temperature results in a darkening of the honey by between 5mm and 24mm Pfund.

Deterioration is pretty severe when only average shade temperatures are considered. But the temperature of containers lying in the sun reaches higher levels. In a recent experiment using a bulk container for shipping, the container on the deck of the ship reached 76.7°C inside. The honey, which was packed in drums in the container, was completely ruined.

Steps which can be taken to reduce the rate of deterioration of colour of honey in storage are as follows:

(a) Use air conditioned storage with temperature below 21°C; or
(b) Use underground storage space; or
(c) Provide well ventilated shade; or
(d) Coat the containers with white paint to reflect the heat of the

sun.

CONCLUSION

It has been shown that by using proper hive equipment, by paying attention to honey extracting equipment and techniques, and by adopting suitable honey processing procedures, deterioration of the colour of honey can be eliminated. Even in storage, damage can be reduced.

References:

SMITH, F.G. (1967) Deterioration of the colour of honey *Journal of Apicultural Research* 6(2): 95-98

WHITE, J.W. et al (1963) How processing and storage affect honey quality *Gleanings* 91(7): 422-425

1964 HONEY HOUSES

There has been a considerable amount of discussion in recent months about changing over from the use of mobile honey houses or extracting caravans to permanent honey houses or central extracting plants. The main reasons given for making such a change are improved facilities for extracting honey and greater comfort, more home life, greater ease of obtaining casual labour and lower cost of operation.

The amateur beekeeper, the part time commercial beekeeper, and the full time beekeeper who lives close to his apiaries would undoubtedly find that having the honey house permanently set up near his home has many advantages. The large scale bee farm, in which four or more men work as a team, would find it most advantageous to have an efficient central extracting plant which would be functioning for a high proportion of working days during the year. But for the commercial beekeeper with 300 to 500 hives, who can manage his apiaries with only one assistant, the issue is not so clear. In this case an efficient mobile extracting van may serve just as well as a central plant. To the owner of an efficient van, the setting up of a central plant means a heavy addition capital expenditure in buildings, equipment and honey supers. Much will depend upon personal circumstances.

We would sound a warning to all beekeepers, and particularly to the one man operator, not to tie up more money than they can afford in buildings and equipment which are not going to be in use for more than a few weeks in the year. It should be remembered that it is the hives of bees, properly managed, which produce the crops. It is more important to have enough colonies of bees to produce a comfortable living than to have a beautiful but unproductive honey house.

1966   CENTRAL EXTRACTING PLANT

To build a permanent central extracting plant or not to build one, that is the question.

For the migratory commercial apiarist, employing one or perhaps two helpers during peak periods and working honey flows more than 240 km from his base, a permanent central extracting plant can be an expensive luxury which is liable to turn him into a road transport driver instead of an apiarist. For him a well designed and properly equipped mobile extracting plant is a more economic proposition.

R.O.B. Manley, operating 2000 hives, considered that 48km was the furthest it was economic to travel from central extracting plant to out-apiaries. But that was in England, with narrow winding roads, small trucks and annual honey crops about one tenth of those obtainable in Western Australia. Miel Carlota, with ten times as many hives and crops approaching ours in size, considered 80km was the maximum economic range in the mountains of Mexico.

Current thought in Western Australia, with our seven-ton trucks and good fast roads, is that 160 to 240km is the furthest it is possible to operate apiaries economically from a central plant. The crucial factor appears to be that if an apiarist can leave home in the morning with a load of empty supers and return home in the evening with a load of 240 full supers, then a central plant is a practical proposition. But if he has to sleep out because of the distance to be travelled or because of the time taken to remove the full supers from the hives, then central extracting is not economic for the average commercial beekeeper.

Ideally, a central extracting plant should have two men fully employed extracting every day. Under our conditions, where major honey flows occur 320km to 640km away from base, at least two teams of apiarists, each with a truck, would be required to keep the central plant fully supplied with supers to extract. This would need 2400 productive hives of bees to employ the plant at full capacity during a honey flow.

A practical compromise for the normal commercial apiarist would be to have an efficient mobile plant so that he can extract near his apiaries when they are out at the more distant honey flows, and to build at his base a skeleton honey house with the additional facilities of a central plant. Then, when his apiaries are close to home, he can set up his mobile extracting van in the skeleton honey house and so obtain the advantages of central extracting.

CHAPTER 13
REFINEMENTS IN THE HONEY HOUSE

While steam has long had its place for heating the uncapping knife, its impact on the quality of honey and wax in this role is so small as not to have caused any concern. But the use of steam in the cappings reducer and in the extractor is another matter; the damage being done was massive. So I turned my attention to encouraging beekeepers to convert their plant from steam heating to hot water systems. Two or three of the innovators among the commercial beekeepers have tried it and are very satisfied with the results.

1967   HOT WATER CIRCULATING SYSTEM

Advantages of the Hot Water System

For many years steam boilers of various types and sizes have been used to provide the heat needed in honey extracting plants. Prolonged contact with the very hot surfaces heated by steam circulating at 100°C or more damages both honey and wax. In addition, the high temperature of the system made working in the plant uncomfortable, particularly in summer when most of the extracting was done. Because there was rarely any arrangement for recovering the steam, the consumption of water was heavy, increasing costs where every drop of water has to be transported to the extracting site. Because much of the heat produced was dissipated, fuel consumption and therefore costs were greater than they need be.

In these days, when every effort must be made to reduce the cost of honey production and eliminate any damage to the honey and wax to get the best prices, the use of hot water circulating systems has definite advantages.

The temperature of the water in the system need be no greater than that required to provide a sufficient temperature gradient in the cappings reducer to melt the wax. As wax melts at between 62.00°C and 62.50°C, a water temperature of 77°C provides an adequate flow of heat to melt the wax in a good reducer.

Because the water is being recirculated all the time, unused heat is retained in the system. This provides a saving in fuel. It also cuts down the consumption of water, only very little being lost from evaporation in the header tank or air release pipe.

Heating unit

The water is heated either by oil or bottled gas. The heating unit is thermostatically controlled so as to maintain the temperature of water within the range required.

When the temperature reaches the maximum of the range, usually about 88°C, the fuel supply is cut off, leaving only the pilot jet burning. When the temperature of the water falls to the bottom of the range, the fuel is turned on again automatically.

The most economical fuel is heating oil. Bottled gas is a little more expensive (1967), but the gas operated heating units are cheaper, cleaner and possibly more reliable.

Oil fired units have been supplied for honey extracting plants by Heatmasters of Wembley and bottled gas units by Rheem of Fremantle.

Circulating pump

A pump is required to circulate the hot water through the system. A ¾in (18mm) centrifugal pump is the type normally used. This has a 1in (25mm) diameter inlet and a ¾in (18mm) diameter delivery pipe. At the top of the casing a small plug is provided which can be removed to allow any trapped air to escape.

If the plant has electricity, the pump can be driven off a ¼hp electric motor. Such a pump can be obtained complete with the motor.

Alternatively, if the plant is without electricity, the pump can be driven by a pulley with a belt from the extractor motor. The speed at which the pump operates is about 1500rpm and it runs continually for as long as heat is needed in the various appliances.

Honey extractor coils

The most convenient and efficient way of heating honey before settling and straining is while it is running in a thin film down the sides of the honey extractor.

If the coils are placed inside the extractor, they form a ledge which holds back some of the honey and wax particles and causes overheating. It is best if the coils are fitted outside the extractor, at a level just below the bottom of the frame baskets.

The coils which we have found very satisfactory consist of a spiral of three turns round the extractor of ¾in (18mm) OD annealed copper refrigerator tubing. This is available in coils of 19lb to 20lb (8kg) in weight and a single coil provides enough for one extractor with a bit over for making connections.

We tin the coils and solder them round the outside of the extractor using five millimetre copper rod as a spacer between the coils to increase the thermal conductivity between the coils and the extractor walls.

At the top, the exit end of the pipe, a small cock should be fitted for bleeding out any air trapped in the system. At the other end a cock is fitted in the pipe line to regulate the flow of water into the coils.

Temperature tests on the use of heating coils round the outside of the extractor showed that hot water circulating at 70-75°C raised the temperature of cold honey at 16°C to 42-43°C. This is the ideal temperature for straining and settling honey.

Water flow circuit

There are a number of different arrangements for connecting up the various appliances in a hot water system. Some create too great a pressure in the system while others produce a vacuum and suck in air.

After numerous trials we have found that the following system provides an adequate circulation of the water, without subjecting the various items of equipment to excessive stress.

The arrangement is best described by means of a diagram. It will be noted that the static head or pressure of the system is governed by the height of the header tank above the appliances. As extracting equipment is generally designed to operate at a pressure of about 34 kPa, the level of the water in the header tank can be between three and four metres above the appliances.

A.-Air escape tube.
E.-Honey extractor.
H.-Hot water heating unit.
K.-Pipes to uncapping knife.
P.-Circulating pump.
R.-Cappings reducer.
T.-Header tank.

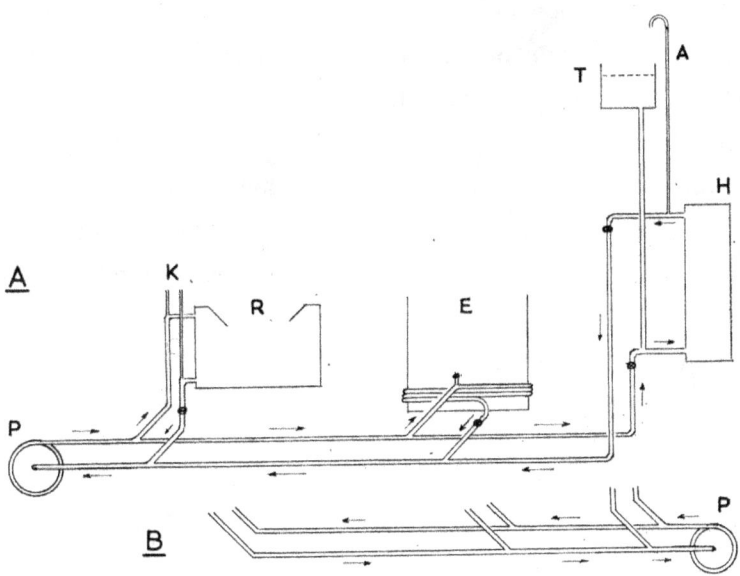

Water Pump Circuit

A Circuit with pump at end of the line, close to cappings reducer.

B Alternative circuit with pump close to heating unit.

The pump is connected in parallel with the appliances and the heating unit. This enables any piece of equipment to be isolated at will. To ensure an even distribution of water circulation, all junctions are of a Y form, the bottom of each Y being towards the inlet or outlet of the pump.

General pipe size should be as large as possible to permit unimpeded water flow; ¾in is a convenient size, with ½in leading to the uncapping knife hoses.

All metal pipes should be lagged or insulated to prevent heat loss. Appliances also should be insulated to retain heat, and to keep working conditions comfortable for the operator. The extra cost of insulation will be amply repaid by increased efficiency of appliances and decreased fuel consumption.

It is important that all appliances have provision for permitting air to escape. This can be achieved by fitting a small cock in the highest position in the hot water system on each appliance. An air escape vent is also fitted near the outlet from the heating unit. This is open at the top end and carried above the header tank. The layout of the pipes should be planned so that no air locks can form in them.

Provision can be made for drawing off hot water for washing down. In this case the capacity of the header tank should be sufficient to replace what is drawn from the system. The header tank should not be allowed to drain dry, otherwise air locks may occur in the appliances and these will have to be bled free of air after the header tank is filled up again.

Plant layout

A hot water circulating system can be used in any extracting plant layout. To conserve heat and economise on fuel, pipe lines should be as short as possible. In an extracting caravan these can be placed beneath the floor.

The heating unit can be located at the front end of the van beside the extractor. This is possibly the most convenient position for the location of fuel tanks outside the van. The header tank can be close to the roof between the extractor and the heating unit.

An alternative arrangement is to have the heating unit on a wheel box.

The pump can be located in any convenient position. If operated from the extractor drive, it can be near the extractor on the wheel box and under the frame rack. If electricity is used the pump and its motor may be most conveniently situated under the cappings reducer. But remember that you need to be able to get at the pump fairly easily to bleed out air should it get into the system.

In conclusion, a hot water circulating system in an extracting plant makes the working conditions more pleasant; enables the temperature at the various appliances to be controlled, reducing the danger of damage to the honey and wax; it saves fuel and economises in water, thereby decreasing operating costs.

1965   THE CAPPINGS

For the small beekeeper, separating beeswax from the honey in the cappings, and rendering the wax into clean cakes is no great problem. But the commercial beekeeper, who extracts 2000 to 4000 kg of honey a day, has to handle a very large quantity of cappings and this can cause some serious problems.

Draining

The simplest method of dealing with cappings is to let them fall into a strainer, and to allow the honey to drain out. In this method the honey is not damaged, but there remains the accumulation of wax to be rendered. The wax can be rendered either by melting it in water and pouring it through a fine strainer straight into the moulds, or by rendering it in one or more solar wax extractors.

The draining method can be used by the commercial beekeeper if he operates a central plant. Miel Carlota, the big beekeeping organization in Mexico, uses it with thousands of hives and a production per hive similar to that obtained in Western Australia. In their case, the cappings fall into a series of hive boxes with wire gauze bottoms; the boxes stand on a long trough which catches the honey draining from them. But this requires more space than a beekeeper can afford in a mobile plant.

Mr K.J. Olley of Oxley, Queensland, has a long stainless steel jacketed trough, in which hang six baskets of fine wire mesh. In a jacket round the trough are electric heating cables, thermostatically controlled. A wooden cover fits over the trough and in one end of the cover is a hole through which the cappings fall from the knife.

As one basket becomes filled it is pushed along the trough and replaced by another basket. During the day while extracting is in progress the trough is heated to only 32°C so the honey drains away without being damaged. When the last basket has been filled, the first basket, from which most of the honey has drained, is filled up again. In this way the six baskets hold a whole day's cappings.

At night, after extracting is finished for the day, the heat is turned up and during the night the wax in the baskets melts, is strained by the wire mesh of the baskets and runs out of the trough with what honey has not drained out during the day. The honey and wax pass through a small honey-wax separator into their respective containers. The wax sets into a perfectly clean cake of high quality wax ready for immediate marketing. The small amount of honey which has been subjected to the high temperature is kept separate from the bulk of the honey and is sold as a lower grade.

In the morning, the propolis which has remained in the baskets, is shaken out and dumped and the appliance is ready for the next day's work.

Centrifuging

To overcome the space factor and to speed up the separation of the honey from the wax, some beekeepers uncap into a centrifugal spin dryer. Factors which effect the efficiency of these are the size of the mesh of the walls of the cylinder, the diameter of the cylinder, and the speed of rotation. Some fine particles of wax may be forced through the cylinder mesh with the honey, so a settling tank is required. From time to time the dry, or partly dry cappings have to be shovelled out of the spinner, to be rendered by some other appliance.

This is not an entirely satisfactory method, and many beekeepers, after giving spin dryers a try, turn to some other method.

Melting

Where space is restricted, the ideal appliance is one which separates the honey from the cappings, melts the cappings and renders them into clean cakes as a continuous process. There is on the market a variety of cappings reducers designed to do this. The main snags in the past are that a fairly large steam boiler was required to provide the heat; the honey tends to be darkened and damaged by overheating; the wax may also be damaged by overheating and may contain slumgum; and the operator has to work under rather hot and uncomfortable conditions.

The essentials of a good reducer are that the honey should be discharged rapidly before it becomes excessively heated; the capping should melt without being overheated and should drain off through a strainer which holds back the slumgum; and the wax should run into a covered and insulated mould in which it sets slowly into a clean cake.

For preference, melting the wax should be done with hot water rather than steam. A hot water generator is cheaper, it makes working conditions less uncomfortable for the operator and hot water is less likely to damage the honey and wax.

None of the cappings reducers commercially available is completely satisfactory, though new designs and improvements are appearing and the Department of Agriculture in Western Australia is working on this problem.

1967   CAPPINGS REDUCER

The modified prototype cappings reducer, for use with circulating hot water, has been in full commercial use at Lancelin and is working most successfully. A few minor improvements have been added to the specifications for future models.

Following the publication in the *American Bee Journal* of an article on the new cappings reducer, there has been a steady demand for plans and specifications. Most requests have came from America (34), followed by Canada (5), Australia (4), New Zealand (2), Mexico (1), and England (1) – 47 in six weeks.

SMITH, F.G. (1966) Cappings reducer: a new design *Amer. Bee J.* 106(9): 333-335

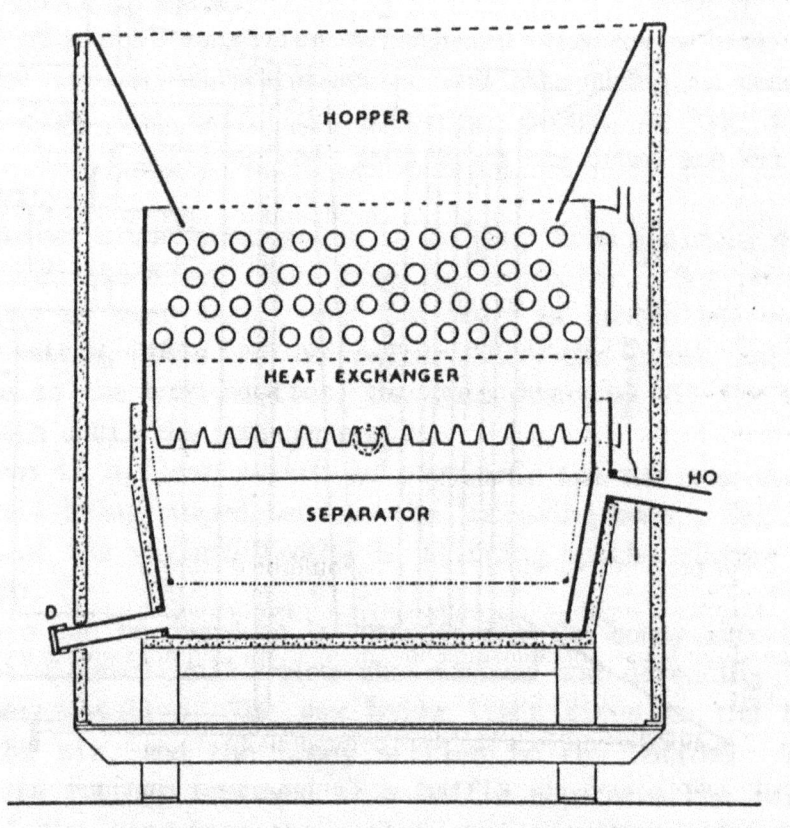

Cappings reducer – cross section

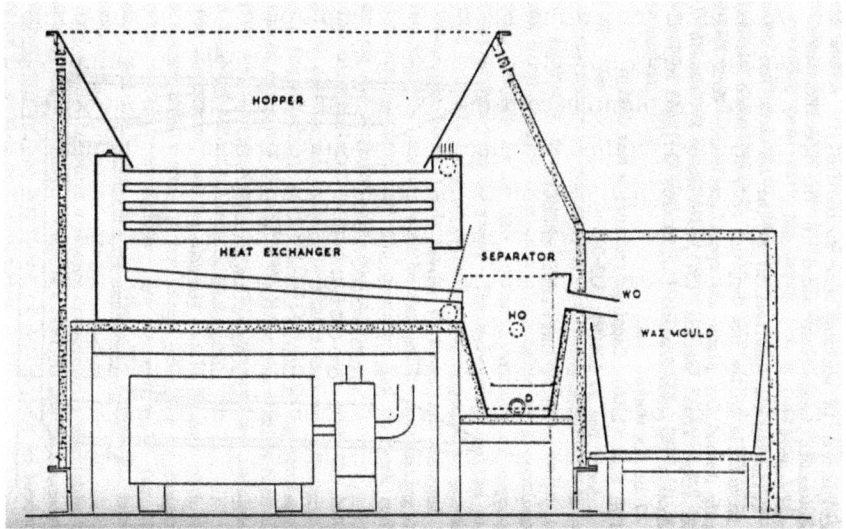

Cappings reducer – longitudinal section

1967   HONEY STRAINER FOR EXTRACTING PLANT

At the last Annual Conference of the Beekeepers' Section of the Farmers' Union, some beekeepers claimed that it was not possible to strain honey through 16 mesh gauze in an extracting plant.

A simple honey sump strainer has been designed for use in mobile extracting plants, and has been tested by a commercial beekeeper. The tests have shown that a much finer strainer than 16 mesh can be used, and further experiments are being conducted to determine the finest mesh that is practicable to use under field conditions.

1968   A HONEY SUMP STRAINER

Honey flowing from an extractor contains particles of wax, air bubbles and small quantities of foreign matter. The removal of this from the honey can be accomplished quite simply in any extracting plant, even in a mobile plant rather short of space.

In the current most primitive arrangement the honey pump is attached to the extractor outlet and everything is pumped direct from the extractor into the drums, or one step better, into a settling tank from which the drums are filled by gravity.

Another crude arrangement is to pump from a simple sump into which the honey flows from the extractor. But unless the sump is very large and the pump is controlled by a float switch, this is no better than the first method because as the sump empties, the pump draws up all the wax and froth until the sump is empty.

Even if a float switch is provided, the wax and froth are still being pumped because the inflowing honey, falling on top of the wax and froth, is stirring up the mixture in the sump.

This can be overcome by introducing the honey into the sump as a level well below the surface and directing the flow horizontally. The wax being light rises to the top with the air, and the honey settles to the bottom. The effect is further improved if a baffle separates the inlet part of the sump from the outlet, thereby maintaining the honey in the inlet chamber at a constant level. The first part of the baffle holds back the froth and wax on and near the surface, while permitting clear honey at the bottom to pass through.

A float switch, to cut out the pump before the pumping section is emptied, will prevent further air and froth being introduced.

The next improvement that can be made is to introduce a strainer. For the preliminary straining, a 16 mesh screen, combined with the baffled settling arrangement, is adequate. By 16 mesh is meant a gauze having 16 wires to one inch measured in both directions.

To work efficiently, the strainer should be immersed in the honey. It should not be held above the surface with the honey dropping through it. This quickly clogs the meshes with wax particles and beats air bubbles in to the honey.

A simple method of straining is to insert a close fitting flat screen between a pair of baffles. This can be changed very easily by putting another screen behind it, and withdrawing the clogged first screen. The screen can then be washed clean with water and dried.

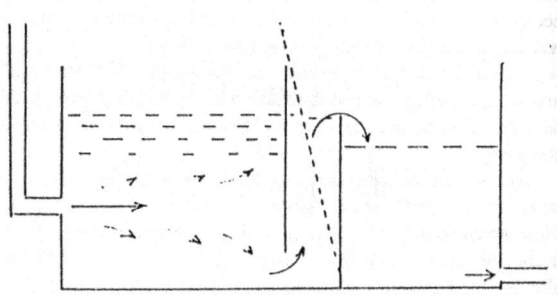

Honey Sump Strainer

For an extracting plant with a larger capacity, or if finer straining is required, the sump can be made with a central straining section containing a basket type strainer. The strainer should hang from the sides of the sump, and be pushed up against the up-stream baffle so that the honey flows over a lip or through a spout at the top of the baffle and into the straining basket.

Between the straining basket and the pumping chamber is a pair of baffles to maintain the level of the honey at below the top of the strainer.

A second basket is kept in readiness. When the first basket begins to become clogged the inflow of honey should be stopped while the clogged basket is raised up over the sump to drain, and the clean basket inserted in its place.

From time to time the accumulated wax and froth in the first or inlet chamber should be skimmed off. This can be put into the cappings reducer.

Provision must be made for draining the sump. This can be done by having a drain plug in the bottom. Additional drain plugs should be provided for the other compartments which are separated by baffle plates.

CHAPTER 14
MAJOR EVENTS

It is appropriate at this stage to consider some of the major events in the Apicultural Section which influenced our work during the life of the journal *Apiculture*.

The impact of the first three issues was demonstrated by the debates in the 1964 Annual conference of the Beekeepers' Section of the Farmers' Union.

One of the younger leading beekeepers, Ken Spurge, opened a discussion on the use of repellents for driving bees out of honey supers. Contributions to the debate were made also by beekeepers from other States; Mr. Hughston from New South Wales and John Guilfoyle from Queensland.

One of Western Australia's most experienced beekeepers, Bill Cook, gave an excellent talk on the disadvantages and advantages of using central extracting plants. He and his brothers had been using a central plant for some years. John Guilfoyle and others made valuable contributions to the debate.

The use of queen excluders was another subject discussed; all speakers advocated their use.

Following the Conference, representatives of the Beekeeper's Section of the Farmers' Union proposed the establishment of standards of proficiency in beekeeping at three levels:

Apiarist's Certificate, the standard of proficiency which an apiarist needs to attain to practice beekeeping in an efficient and economic manner and on a scale which will provide him with a reasonable livelihood.

Senior Apiarist's Certificate, the standard of proficiency for the experienced beekeeper who may wish to aspire to a higher qualification. The examination would be open to candidates who had passed the Apiarist' Certificate examination and have kept and managed bees for at least three full years, and to candidates who are established commercial beekeepers who have been operating more that 200 productive hives for at least five full years, and during that period have been dependent mainly on beekeeping for their livelihood. The standard of the examination would be equivalent at least to that of similar examinations in other countries.

National Diploma in Beekeeping is the standard of proficiency intended for those who may wish to qualify for positions in beekeeping education. In addition to covering the field of the Senior Apiarist's Certificate, it requires a broad scientific knowledge and familiarity with beekeeping literature, recent research, and extension methods. It was established with the approval of the United Kingdom Ministry of Education and Ministry of Agriculture and Fisheries. The examination can be held at approved centres in Western Australia.

It was also proposed that appropriate courses be made available by the Technical Correspondence School, and a practical course be run in conjunction with the Apiarist's Certificate course.

The above was endorsed unanimously by a meeting of representatives of the Beekeepers' Section of the Farmers' Union and the WA Apiarists Society held at the Department of Agriculture on August 17, 1964.

A departmental bulletin on training in beekeeping, which gives full particulars of each of the standards of proficiency, was published soon afterwards.

TOUR OF THE EASTERN STATES

An invitation had been received for me to attend and contribute an address to the Field Day to be held in November 1964 at Glenrowan in north east Victoria. The Beekeepers' Section of the Farmers' Union recommended that I be permitted to confer with the Apiculturist in each of the other States to discuss problems of mutual interest and research, and if necessary, to resolve minor difficulties that may exist between the States. Approval for the tour was given by the Minister for Agriculture and I spent three and a half weeks in Queensland, New South Wales, South Australia and Victoria, conducted by the apicultural staff in each of those States.

The tour enabled me to see what the most progressive beekeepers in Australia were doing and how they had overcome the very problems about which I was so concerned in Western Australia. Further it resulted in a spirit of sympathy and comradeship between all of us who were employed by the State governments, and a comprehension of where our real problems lay. I produced a report on my tour of the Eastern States which I distributed to all beekeepers in Western Australia.

In a later chapter I will enlarge on the things I saw on the tour which I believed would be of benefit to Western Australian beekeepers.

ADDITIONS TO APICULTURAL SECTION STAFF

For the financial year 1964-65 I obtained authority to employ an additional Apiary Inspector and Bob Coleman took long leave to study dentistry and did not return.

In January 1965 we had the pleasure of welcoming two new members of the Apicultural Section staff, Mr Stan Chambers and Mr Lee Allan.

Stan Chambers was no stranger to many beekeepers in Western Australia as he had served with the Apicultural Section for seven years in the 1950s. His past experience enabled him to take on the full load of his duties as soon as he had become acquainted with the changes which had taken place since he left the Department. He assisted Alan Kessell in carrying out the regulatory duties, field extension and advisory work of the Section. During the past year we had been too short-handed to do much more than attend to reported outbreaks of disease, complaints about bees and neglected apiaries. But from then on, beekeepers would get more visits as the inspectors worked through their respective districts. As far as possible, one of them would always be in the office to attend to queries while the other took his turn in the field. Already, Stan Chambers had taken a most active part in the organisation of the highly successful Field Day held on 6th March.

Lee Allan came to us from Muresk where he had obtained his Diploma in Agriculture. In spite of his youth, he had been a beekeeper for eight years. He would be engaged mainly in assisting me in the research work, both in the laboratory and in the field. Already he had made substantial inroads into the backlog of routine laboratory preparations.

We expected to be able to give beekeepers in Western Australia much more of the help they needed.

FIELD DAY

The Beekeepers' Section of the Farmers' Union and the W.A. Apiarist' Society both decided that they would like to have a field day. This was organised by Stan Chambers and held on 6th March 1965 at the Wattle Grove property of Bill Carvill. The number of beekeepers and their friends who attended greatly exceeded the expected turn-out. The field day consisted entirely of demonstrations and displays of bees, equipment and techniques – there was no room for time-filling speeches, talks, lectures or addresses, other than a very brief opening and closing of proceedings.

On display were hives with 286mm deep brood frames containing the brood nest in one 10-frame box (Jumbo), special honey-extracting supers (Manley) which are easy to handle, improved designs of hive lids and bottom boards, baby nuclei for queen raising, and Sid Murdock's multi-queen coffin hives.

Bill Cook of Toodyay demonstrated his techniques of commercial queen raising. The three main races of honey bees, Italian, Carniolan and Caucasian, were demonstrated so that their characteristics could be readily compared.

The uncapping machines produced by Pender Brothers and the Root Company were demonstrated – unfortunately the Guilfoyle machine did not arrive in time – and an opportunity was given for beginners to try their hand at honey extracting. An efficient new type of solar wax reducer, particularly good for rendering cappings, was on display. A picture display of the Bee Research Headquarters at work and the services provided by the Department of Agriculture attracted considerable attention.

The various kinds of equipment used for loading and moving apiaries were demonstrated, with the numerous boom and gantry loaders used in this highly mechanised State forming a spectacular display. Visitors also had the opportunity of inspecting a selection of honey extracting caravans.

The day concluded with a smoker-lighting competition which had to be run in a number of heats, with Tom Powell, Manager of the Honey Pool, reaching the final with his highly entertaining and unique technique. Unfortunately, his smoker would not shut in the final.

The proceedings filled a thoroughly enjoyable day from 10 in the morning to 4.30 in the afternoon, and attracted beekeepers from all over the State. And for those who could not get to it, both TV stations provided fine coverage with the ABC's Tony Parker getting ten stings on the head when trying to record the sounds of a colony of bees. I have a feeling that that episode marked the abrupt end of my popularity with the ABC.

* * *

The following are editorials reporting major events, which I published in *Apiculture*:-

1965   VISIT OF PROFESSOR MYUKOLA HAYDAK TO WA

We received a most enjoyable and instructive visit from Professor Haydak, professor of apiculture at Minnesota University, and a world authority on the nutrition of honeybees. Professor Haydak, who was accompanied by his wife, had been working at the Waite Institute, Adelaide, for the previous three months. While there he adapted to Australian conditions his technique for determining the nutritional value of substitutes for pollen.

A record audience of all sections of the beekeeping industry attended at the Department of Agriculture's lecture theatre at South Perth to hear the Professor lecture on the nutritional requirements of the honeybee. He explained the subject in a most lucid and charming manner which enabled all to understand and appreciate his points. After the formal lecture and questions, the meeting continued informally in the canteen over supper.

1965 ECONOMICS OF BEEKEEPING

In response to enquiries, a bulletin had been prepared on the economics of beekeeping in Western Australia from the honey producer's point of view. This was examined before publication by several leading commercial beekeepers to ensure that it presented an accurate picture of beekeeping economics of the day under local conditions. It was hoped that when published it would provide a useful basis for calculation of the profits of honey production so that beekeepers could get a true and clear picture of how they stood.

By the summer of 1966, the bulletin had been published and was in the hands of all beekeepers in Western Australia. They were recommended to read the bulletin carefully and to consider their own position in the light of the information it contained. It was concluded that with the price of honey at its current level, there could be very few beekeepers whose bees were keeping them.

1966 AWARD OF NATIONAL DIPLOMA IN BEEKEEPING

Mr Alan Kessell was successful in passing the examinations for the National Diploma in Beekeeping.

Acting on behalf of the Examination Board for the National Diploma in Beekeeping, Mr. L.A. Logan, Minister for Local Government, publicly presented the Diploma to Mr Kessell at the opening of the Annual Conference of the Beekeepers' Section, Farmers' Union, on 5th July 1965.

THREE CELLS OF HONEYCOMB

1966   OPEN DAY AT THE DEPARTMENT OF AGRICULTURE

As the 1966 annual Conference of the Beekeepers' Section of the Farmers' Union coincided with the occupation of the new apicultural laboratory, special arrangements were made for the members of the Beekeepers' Section to attend an open day at the Department.

In the comfort of the Department's theatrette the commercial apiarists were given a short talk on the Dadant depth honey supers with Manley frames and were shown a film on the production of beeswax and honey in Tanganyika, a type of beekeeping very different from that practised in Australia.

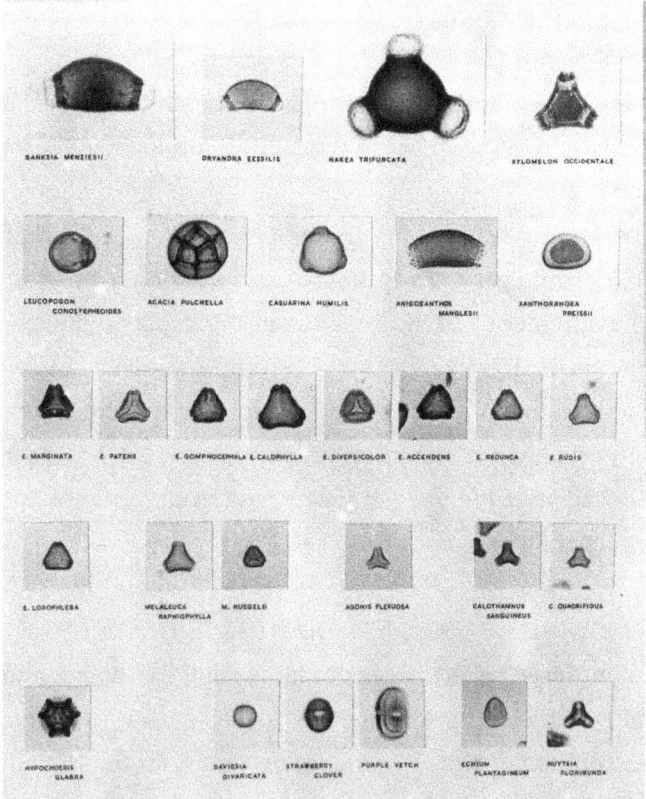

Following this, the beekeepers were able to see a display of the various laboratory activities performed in the Apicultural Section. The bee forage studies were illustrated with a collection of Eucalyptus from the Goldfields and east of Hyden, Dr Beard's vegetation maps, the description and photography of pollen grains and the identification of pollen grains in honey and pollen loads.

There was a demonstration of the preparation of pollen grains for microscopic examination, together with honey analysis procedures, particularly in respect of pollen analysis, colour grading, determination of water content, and the measurement of reducing sugars and sucrose. Also on display was a map showing the extent of the big fire in the Banksia and Dryandra areas at Lancelin.

The final phase of the open day took place in the processing room in the bee products building, where the visitors saw honey extracting equipment from various parts of the world under test, including the prototype of our own cappings reducer operated by hot water and the new Hills uncapping machine kindly provided by Mr John Guilfoyle.

The open day was attended by almost the entire membership of the Beekeepers' Section of the Farmers' Union, as well as by some visitors. The displays gave the apiarists an opportunity to see and appreciate a part of the work that goes on in the Department of Agriculture's Apicultural section.

1967   HONEY QUALITY STANDARDS

Following the Open Day, the *Government Gazette* of 27th October 1966 announced the Regulations under the Beekeepers Act governing the standards of quality with which all honey offered for sale in Western Australia must comply. The quality standards apply to honey sold by producers to packers and to honey sold to the public. The standards also cover packaging. The quality standards are very simple and present no difficulty to the producer or packer who practices common sense care in the handling of his honey. The aspects covered are the degree of ripeness or water content as indicated by the refractive index, the cleanliness of the honey, freedom from damage and freedom from adulteration. The details given in the Regulations were published in *Apiculture*.

1967   VISIT BY PROFESSOR ERDTMAN

Professor G. Erdtman of Stockholm, author of several text books on pollen morphology and editor of *Grana Palynologica*, spent three weeks in Western Australia accompanied by his wife and Professor J. Rowley of Massachusetts.

Arrangements were made for the party to see as much as possible of the local flora in the time available, and to meet all who were interested in their work and who could help them in their search for material.

Professor Erdtman and his wife worked for much of their time in the State Herbarium, while Professor Rowley prepared his material for electron microscopy in the Apicultural Laboratory.

1967   THE HIVE – A NEW BULLETIN

A new bulletin was published by the Department of Agriculture, Western Australia, *The Hive*. It is the most comprehensive work ever published on this subject and is fully illustrated with photographs and line drawings. The bulletin contains the latest information on the most recent developments in hive design, recommendations concerning the most efficient equipment, and instructions for assembling hives ready for use.

The bulletin is written in a clear and lucid manner, and is extremely well printed on good quality paper with graphs, 25 line drawings and 3 tables of dimensions.

1967   APPOINTMENT TO APIMONDIA COMMISSION

Dr Francis G. Smith was elected a member of the Commission on Bee Technology and Equipment of Apimondia, the International Federation of Beekeepers' Associations.

Announcing this, Professor V. Harnaj, President of Apimondia, stated that this appointment was in recognition of Dr Smith's competence in this specialised field and of his steady work towards promoting world apiculture.

Earlier, Dr Smith had been invited to present a paper during the proceedings of the Bee Technology and Equipment Commission at the 21st International Beekeeping Congress held at Maryland, USA. Unfortunately, approval to attend was not granted and he was obliged to decline the invitation.

1967   DR EVA CRANE'S VISIT

Dr Crane, Director of the Bee Research Association, ended her tour of Australia by visiting the beekeeping areas of the West.

She was met at Perth airport on Saturday, 18th November by Mr Jack Knight, President of the Beekeepers' Section, Farmers' Union, Mr Ken Healy, Vice President, Mr and Mrs Tom Powell, Mr R.E. Smith and Dr and Mrs F.G. Smith.

The official programme began on Monday with paying official calls, and by giving a mid-day lecture at the Department of Agriculture on 'Scientific Papers for Publication; for the point of view of author, referee, editor, publisher and abstractor'.

During the afternoon, Dr Crane visited the honey packing plants of the WA Honey Pool and Sunny Flo Apiaries. The next two days were spent on a tour of the South West Jarrah and Karri Forest areas. We were able to show her a very good cross section of the beekeeping industry and the flora, and she was able to see apiarists at work in their apiaries and extracting plants and in a queen-raising establishment.

On the Friday evening, a very large attendance of beekeepers was present for her final lecture on 'Bee research in the world today and how it can help the Australian honey industry.' Dr Crane pointed out that while our flora, honey and pollen were different from those in other parts of the world, and needed local research work, the fundamental research carried out on bees and their diseases in Europe was applicable everywhere. It was therefore important to concentrate our resources in research on what can be studied only here, and to make use of the work done elsewhere in the world on subjects of universal application.

Dr Crane commented on the lack of knowledge in Australia of research which had been done elsewhere and the lack of means of ready reference to such research. She suggested means by which this could be overcome, through using the facilities provided by the Bee Research Association.

Dr Crane remarked on the excellence of the different kinds of honey as produced by the bees in Western Australia, and the unfortunate effects of some beekeeping methods and of blending different qualities of honey. She pointed out some of the aspects in which we needed research, particularly on our honey which had a great variety of qualities awaiting chemical research, and which needed a thorough exploration of all means whereby it can be used.

It is clear that Dr Crane's visit has been a great success from all points of view, and her talks were the best that have ever been addressed to the beekeeping industry in Western Australia. We will be looking forward to reading the account of her impressions in *Bee World*.

1967-68   OUR HONEY AND DR CRANE'S REPORT

Our honey is the subject which has claimed most of my attention during the past five years. I believe concentration on this is the most important matter in our industry. It is the one thing which can get us out of our present difficulties and enable us to establish ourselves on a sound basis for the future.

Dr Crane's visit to Australia and her subsequent report has been a highlight of the whole period, and has given direction and purpose to our efforts, not only in Western Australia, but throughout the Commonwealth.

She considered that bee management in Australia was at or above present world standards, but that honey management needed urgent and drastic attention. She believed that the present difficulties in getting a satisfactory price for exported honey are linked with the way honey is treated here in Australia.

Dr Crane suggested that the most important function of the Honey Board with regard to honey offered for export should be to improve quality. She believed that the quality of many of the different types of honey produced by the bees and extracted by many of the beekeepers was perfectly acceptable to overseas markets, but that various forms of deterioration can be induced in handling.

She recommended that the characteristics of Australian honeys be studied, and that changes induced in flavours during handling and processing be investigated.

These and other recommendations made by Dr Crane to the Honey Research Advisory Committee received the unanimous support of the State Apiculturists at their meeting in Canberra in March 1968, and were accepted by the Honey Research Advisory Committee.

As a result I was asked by the Committee to contact British and German honey chemists to obtain information on the technical side of the trade, with a view to obtaining the quickest improvement in the acceptability of Australian honey overseas. I was able to report back to the Department of Primary Industry and the Honey Research Advisory Committee at the end of May, and we now know precisely what our overseas customers require.

The requirements form a long list, but basically they amount to honey as produced by good apiarists, packed in the type of container specified by the customer.

Packers who have developed processing and blending techniques to dispose of all honey to mass markets when there was a shortage of honey, may find the requirements difficult, but they are not impossible.

Today, overseas buyers can be selective in their choice of honey. If Western Australia does not supply what they need, they will buy elsewhere. It is not wise to forget the old trading principle, "**the customer is always righ**t".

1969   UK HONEY MARKET

While this is hardly an event of our own it is appropriate to look at what others are saying about our efforts, so I include the following from the *Financial Times*:-

"To compensate for the loss of manufacturing outlets in the UK to the Chinese, the Australians are putting more emphasis on the quality end of the trade, thus helping to widen the division in the UK between industrial and table markets.

"Hitherto, Australia has mostly lumped all its different kinds of honey in together and sold the result by colour, but producers are now being urged to keep them separate so that they can be marketed on the more sophisticated basis of floral sources, attracting an additional quality premium of perhaps £5 or £10 a ton, adding 2d or 3d a pound in shops for special kinds of honey, but still far short of the prices commanded by the more exotic Continental types, such as Greek.

"In this shifting of emphasis, Australia has been helped by the decline in the production of its manufacturing-type of honey in the past year – partly because of seasonal conditions and partly as a result of reforms aimed at pulling the industry out of its troubles by exploiting the country's floral richness and adopting a tighter marketing system. After some false starts, the Australian Honey Board seems to have learned how to attune themselves to the market.

"The next objective is to leave manufacturing sales to China and consolidate at the more prosperous, if not quite luxury, end of the market. Time may thus show China to have done a service to the standards of the British honey trade.

"Honey prices, which have previously been higher than they are now in times of uncertainty, are likely to stay up until the end of the year, when levels will be decided by prospects for the new season's supplies from major sources – Australia, Argentina and Mexico. Whether they go higher between now and then depends partly on how world stocks, at present tight, are affected by buying from the Continent, where some reserves are running low.

"Latest figures suggest that British honey consumption is beginning to level out from the steep rise of the past few years, so if slow-down coincides with heavy yields when buying begins for the next year, the market could revert to lower levels.

"Traditional suppliers would then have reason to be thankful that the effects of China's honey shipment are limited. Although good for manufacturing, Chinese honey does not bend well and has been turned down for table use because of its high iron content."

1970   HONEY EXTRACTION

Recently a bee journal in Britain carried a picture of a woman putting a comb of brood into a honey extractor. The caption of this picture was 'Extraction in Australia'

In an even more recent bee journal in Australia we see a comment that one apiary has had the brood combs extracted to give the queen room to lay. This was passed over in the article concerned as being quite a normal practice.

Are we trying to produce honey for human consumption, food unique for its flavour and aroma and other properties? Or are we just trying to produce something to sell, not caring what happens to it after it has left our hands?

Nowadays the majority of beekeepers are very much concerned about quality, and fully appreciate the need to maintain the quality of the honey as produced and stored by the bees, and to do nothing that will in any way spoil the honey. But unhappily, old habits and old ways of thinking are very difficult to get rid of. The idea that it is necessary to extract honey from brood combs to provide room for the queen to lay is one that dies hard. The proper way of providing room in a hive that is congested is to give another super or to extract the supers that are full of ripe honey. If the bees want to expand their brood nest, they will move the honey and pollen from the comb do to so. They will in fact, use most of that honey and pollen to feed the brood.

Another point to remember is that if you rob the brood nest – and this is robbing the bees – you are depriving them of their reserves of food. This can lead to demoralization and loss of spirit in the colony. If the weather changes or the honey flow ceases, it can lead to starvation, chilling in winter, or collapse of a colony from heat in the height of summer. It must be remembered that reserves of honey in the brood nest act not only as emergency food supplies, but also as a buffer against rapid changes in the temperature outside the hive.

The Codex Alimentarius makes it quite clear that honey is to be extracted only from combs that are sealed and which do not contain any brood.

CHAPTER 15
EASTERN STATES TOUR

I was interested conferring with the Apiculturists in each of the other States, studying research work in progress, visiting leading beekeepers and equipment manufacturers and learning what the beekeepers were doing. In particular I wanted to see the designs of hives in use and the methods of apiary management and honey and beeswax extraction. These were aspects which might be of use to beekeepers in Western Australia. In this digest of that tour I have concentrated on those matters which I considered to be most relevant to our problems in WA.

HIVES

In Queensland the usual hive is the 10-frame Langstroth. Where lightness in the weight of the individual box is required, the 'WSP' is popular. This box is 50mm shallower that the Langstroth.

In New South Wales, Langstroth 10-frame, Langstroth 8-frame, WSP and even Langstroth 12-frame hives (which they call Jumbo) were seen to be used by commercial beekeepers.

In Victoria, the 8-frame Langstroth is almost universal. However there are some 10-frame Langstroth hives, and this size is increasing in popularity.

In South Australia the 10-frame Langstroth is normal, but C.S. Scarfe, who is renowned for his readiness to try out new ideas, is experimenting with long boxes containing 5 queens, and with the use as supers, shallow boxes known as 'Ideals' with frames 136.5mm deep.

Queen excluders

In Queensland, queen excluders are in universal use. The metal bound wire excluders are considered to be the best. Queen excluders are also widely used in New South Wales. In Victoria queen excluders are not generally used, but there is an increasing interest in their use. They are not used in South Australia.

Hive lids

The most widely used hive lids in Queensland, Victoria and South Australia are the ventilated migratory type. In New South Wales they tend to be unventilated. With ordinary migratory lids, a plastic mat, somewhat smaller than the inside of the hive, is used as an inner cover resting on the top bars of the frames. The bees stick this down with propolis.

In South Australia large numbers of the American type of metal covered telescopic lids were seen, but C.S. Scarfe was using a simply shade board lid, consisting of two sets of 19mm thick boards, separated by three cleats to provide 13mm or so air space between the top and the flat lid. These are similar to the type which earlier was common in Western Australia, but was abandoned as it was too good a nesting place for re-backed spiders.

In common use in Queensland were ventilated migratory lids with built-in inner covers. A 13mm gap is left on each side of the inner cover to permit air to pass from within the hives to the exterior through the lid ventilation holes. These are similar to the migratory lids developed by the Department of Agriculture in Western Australia.

Bottom boards

In general these are of the normal Australian type, being of wood or tempered hardboard, with risers to provide bee space under the frames and with cleats underneath.

The exception is Queensland where several types are used. Some have ventilated metal bottoms, all have cleats underneath. The entrances are small, with metal shutters or pivoted wood closures. Most of the bottom boards seen have a baffle or false floor, like the built-in inner cover in the lids, with a 13mm space along each side, and with bee space above the baffle. K. Mitchell of Warwick and also C.S. Scarfe in South Australia are using the old Miller type of bottom board with the removable frame of slats and a 100mm baffle near the entrance. This type of bottom board has been given some publicity in recent years in America by C. Killion, who uses it in the production of comb honey in sections.

The advantages claimed for the baffled bottom board and for the Miller type are that they prevent the combs from being cut away by the bees near the entrance, and they provide clustering space under the combs in hot weather.

HIVE MANAGEMENT

It is clear that hive management has reached a high level of efficiency in Queensland and New South Wales. Adequate numbers of supers are used, with queen excluders to confine the brood nest in the bottom box. The concept of harvesting, removing boxes full of ripe honey, is widespread.

In Victoria and particularly so in South Australia, the harvesting concept is not well developed, and robbing hives of their honey is more normal.

Moving apiaries

In the Eastern States there is very little moving of hives with open entrances. All hives are fitted with entrance shutters, usually a strip of metal pivoted on a mail at one end, with a bent nail at the other end to act as a catch.

The most popular type of fastening to prevent the parts of the hive from moving in transit is metal strapping, secured by Emlocks. Galvanised strapping is used by some beekeepers. Some strapping machines are used in Victoria. Some more simple and cheaper fasteners for strapping were seen, but they appeared less convenient to use than Emlocks.

Wedge Clips were in evidence in Queensland, New South Wales and

Victoria, and in Queensland they were seen on the sides of the hives, with flat wooden spacers attached to the sides of the migratory lids.

Spring clips are normal in South Australia and were also seen in Victoria. The old flat type of South Australian spring clip is being replaced by the bent wire type which is used in Western Australia.

Travelling screens seem to gone out of use.

Hive loading systems

The boom loaders and gentry loaders used in Western Australia were conspicuously absent in the Eastern States.

There are a few Kelly boom loaders in New South Wales and these are considered to be more efficient and faster in operation that the WA boom loader. But even so, booms are considered to be too slow and too expensive. The opinion was expressed that there is a need for a more efficient boom loader costing no more than £500.

The most popular mechanical method of loading hives in Queensland, New south Wales and Victoria is by means of a two wheeled hand barrow, which is run up a ramp on to the truck deck.

Light weight barrows with bicycle wheels were seen, and with one of these to each man, two men can load an apiary in one hour.

Motorised hand barrows were well in evidence, with a clutch to each wheel. These appeared highly satisfactory in taking the heavy work out of loading.

For ease of loading with a hand barrow, and for full utilization of the truck deck, provision was made for the tail gate to be held in a horizontal position as a loading platform.

In South Australia and Victoria there are occasional tractors fitted with fork lifts for loading hives. These tractors are towed or carried on a trailer behind the truck. They can also be used for clearing and levelling apiary sites.

Some small motor operated hoists with miniature booms were seen, but were not particularly efficient.

HONEY EXTRACTING

Central Extracting Plants are very much in evidence in Queensland and New South Wales. There are none in Victoria and only one, Scarfe's, was seen in South Australia.

Uncapping is normally by the standard steam heated knife. The mounted steam-heated vibrating knife is rare, but users claim that after the initial stages of getting combs in good order, the rate of uncapping is twice as fast as with the ordinary steam knife. The fact that both hands can be used for holding the frames of honey is a great help.

The Fox Harrison uncapping machine is used by Murray Charlton in NSW and is being installed by Scarfe in South Australia. However, as it can uncap the equivalent of two truck loads of supers of honey in an eight-hour day and is very expensive, it is a tool which can be used to advantage only by the most extensive beekeeping operations.

Cappings melters and other devices for separating the honey from the cappings and rendering the cappings into cakes of clean wax are, in the Eastern States as in Western Australia, the weakest point in the honey extracting procedure.

None of the cappings reducers was really satisfactory, though there were some quite good home made designs. The best designs separate the honey from the wax before the wax is melted, and heat the wax with hot water, not with steam.

The most effective device seen, which was suitable for central extracting plants, was that made by K.J. Olley of Oxley, Queensland, and has been described in Chapter 13.

The honey extractors seen in Queensland and New South Wales showed a great improvement on the traditional design. Made with tanks of stainless steel, and with the iron parts finished in pale grey enamel, they are hygienic in appearance and a great improvement on the old black and red painted and galvanised iron machinery. Stainless steel lids are provided to cover the tank when the extractor is in motion. The bottom bearing, which has long been a cause for complaint, and which can only be reached by turning the extractor upside down, or by having a trap-door in the floor, is now fitted with provision for greasing.

The semi-radial type of extractor is most commonly used in the 9, 12 and 21-frame sizes, but Murray Charlton has two large imported radial machines which he finds highly satisfactory. The key to success in the design of a radial extractor is to have a large diameter tank, so that there is the minimum of difference in the centrifugal force applied to different parts of the combs, and to have proper gearing so that the combs accelerate slowly, extracting the bulk of the honey at low speed before drying off the combs at high speed. These features are lacking in Australian made extractors.

A device which allows the semi-radial extractor to run for a set time in one direction, reverse the direction of drive and then stops the machine at the correct time, increases the efficiency of the honey extracting process.

This has been achieved by Scarfe by a very simple inexpensive mechanical system of levers operated by a worm gear on the top of the rotor shaft.

Straining and settling honey is normally carried out in Queensland by straining in a honey sump followed by settling in a large storage tank. K. Mitchell in Warwick uses a series of large connected tanks with baffles, all enclosed in a case which has electric heaters under the tanks. The weather can turn very cold in Warwick. Scarfe has a large baffle tank.

Warm rooms for heating honey in the supers were under construction. Some had heaters with fans, mounted on the wall and circulating heated air through the room. These were usually heated by hot water. In one case air in a duct was heated by an automatic oil burner and blown into the room at ceiling level, cold air being drawn out through ducts near the floor. Other warm rooms are heated from close to the floor or from under it. In one case black electric heating elements were fitted round the bottom of the walls.

CONCLUSIONS

Hives

The need for shallower boxes, particularly for use as honey supers, was shown by the use of WSPs in Queensland and New South Wales and of Ideals by the innovative C.S. Scarfe in South Australia. The use of 8-frame Langstroth boxes in Victoria and to a lesser extent in New South Wales demonstrated the desire for lighter boxes for honey supers. The 12-frame Langstroth hives were used by some commercial beekeepers who recognised that a single 10-Lanstroth box was too small to accommodate the brood nest at some seasons, while two standard Langstroth boxes was too much.

The ventilated hive lid with built-in inner cover seemed the most useful type. It saved having first to take off the lid and then having to prize off the propolised inner cover. The operation of opening the hive was done in a single motion with least disruption to the bees. Further, the built-in inner cover with the slots at the side stopped the propolising of the wire on the inside of the vents situated at the ends of the lid.

The use of baffled-entrance bottom boards showed a distinct advantage by ensuring full use of the comb area available in the brood box, as well as providing more clustering space in the hive. There appeared some scope for design development to reduce the cost of manufacture of these boards.

Hive management

The use of metal strapping secured by Emlocks was by far the most convenient. The initial cost of Emlocks is offset against the hire of strapping machines and consumption of clips and strapping. With Emlocks, the strapping is reusable.

A totally satisfactory loading system is yet to be evolved. The WA boom loaders powered by second-hand aero-engine starter motors are practical but slow and the Kelly boom loader, being imported, is also expensive. There is scope for inventive genius here, hydraulics combined with springs seems to offer prospects for a fast working loader.

Honey extracting

The mounted vibrating uncapping knife seems the most useful advance on the hand-held knife, but the equipment for handling the cappings without damage requires development.

The modern stainless steel extractors with improved bottom bearing and, for central plants the large diameter radials, show a big step forward.

*   *   *

The full text of the Report of an Eastern States Tour was published and distributed in 1965.

CHAPTER 16
HIVES

In this chapter I have collected the articles I wrote about aspects of hive design which I considered to be of particular relevance to commercial beekeepers in Western Australia.

THE HIVE AS A TOOL

The hive has been defined as a man-made nesting place for bees. The requirements of a nesting place for bees include being a cavity of sufficient size to accommodate the fully developed bee colony, providing protection from extremes of heat and cold, from winds and from rain, and being defensible against natural enemies. In addition to being a good nesting place for bees, the hive must also meet the requirements of the beekeeper.

The hive must be strong enough to withstand fairly hard usage, be light enough for handling, resistant to fungal decay and insect attack, and to be reasonable in the cost of construction.

The commercial beekeeper requires his hives when occupied by bees to be easy to move without killing the bees or damaging their combs. The hives must be easily adjustable in size according to the season, convenient for the collection of the honey crop with the minimum of disturbance to the bees, and with combs, particularly brood combs, readily accessible for examination.

It cannot be assumed that any particular type of hive produces bigger crops. It is not the hive, but the nature of the bees in the hive, the management of the hive by the beekeeper and the availability of nectar-producing plants that make the difference.

A well designed and accurately made hive adds to the efficiency and pleasure of beekeeping and helps to make beekeeping a productive and profitable occupation.

THE BROOD BOX

The two writers on beekeeping who had impressed me the most were William Hamilton and R.O.B. Manley. But their approaches were entirely different; the former using two boxes for brood nest management during the build-up season and the latter a single large box throughout the year. My experience of beekeeping in the UK made me appreciate the doubling system for use in Scotland and the north of England. But in the south of England, in Africa and again in Australia the single large brood box method of management appealed, not only for simplicity of management, but also for cost – only one box of brood frames being needed, not two.

Some of our own beekeepers in Western Australia had found that a single Langstroth 10-frame box was too small for the brood nest at times; this had been stated openly at the Annual Conference of the Beekeepers' Section of the Farmers' Union. This was also appreciated in New South Wales where some beekeepers had adopted a larger single brood box containing 12 Langstroth frames, increasing the area available for the brood nest by 20 per cent.

Langstroth patented his 10-frame hive in 1851. Early advocates of larger hives were Quinby and the Dadants. After many earlier experiments and trials, the Modified Dadant hive was brought out in the USA in 1920. It contained eleven frames of Quinby depth, 11¼ inches (286mm) and of the Langstroth length, 17 5/8 inches (447mm). The frames were spaced at 1½ inches (38mm). My own experience with the Modified Dadant hive led me to believe that this hive, which held eleven of the extra deep frames, was larger than was needed for the brood nest under the conditions I had met; it is 37.3 percent bigger than the 10-frame Langstroth. R.O.B. Manley had the same experience and as a result he used only ten frames in the brood nests of his Modified Dadant hives.

Shallow supers always are used with the Modified Dadant hive and are added above the brood nest to contain the surplus honey gathered by the bees. These supers are 6 5/8 inches (168mm) deep and take frames 6¼ inches (159mm) deep.

The Jumbo Hive was sometimes called the Quinby hive, but the latter had slightly longer frames. The Jumbo hive has the same horizontal dimensions as the 10-frame Langstroth, and holds ten frames of the Quinby depth and the Langstroth length, spaced at 1 3/8 inches (35mm), the same as in the Langstroth hive. It is of the same depth as the Modified Dadant; the frames being 54mm deeper than the Langstroth. This increases the area available for the brood nest by 25 per cent, which in practice seemed about right, and is of the same brood nest capacity as the Modified Dadant with ten frames. I came down in favour of the Jumbo, as being the brood box most simple to use and easy to handle. And it had the advantage of being compatible with the rest of standard 10-frame Langstroth equipment.

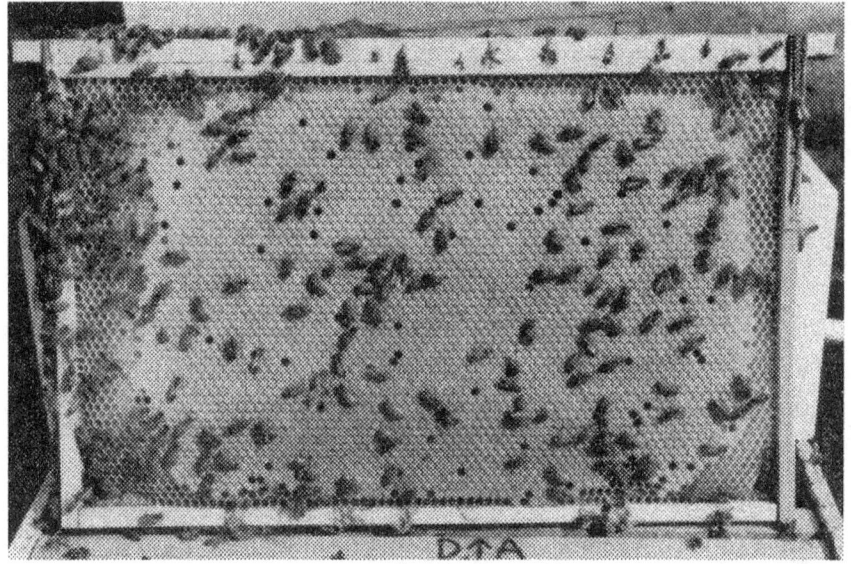

Modified Dadant/Jumbo brood frame

In the Jumbo hive, the brood nest is contained in the bottom box throughout the year. In my experience ten of the 286mm deep frames provide enough room during the build-up and there is ample room for food reserves. There is a great saving of time in managing Jumbo hives. Honey is taken off quickly from above the queen excluders and brood nests can be checked very rapidly. The large uninterrupted combs provide more natural conditions for the brood nest and egg-laying can proceed at the maximum rate. The brood being on a smaller number of large combs, it is much easier to assess the characteristics of the queen and the condition of the colony. The presence of disease is more readily detected.

Having 25 per cent more comb area, the Jumbo brood box is heavier than a Langstroth box, but it is lifted only when moving the whole hive to another site. There is little or no difference in weight between a two-box Langstroth hive and a Jumbo hive with one shallow honey super. A three-box Langstroth hive is distinctly heavier than a Jumbo with two shallow supers.

The 10-frame Jumbo was used and strongly recommended by the 50,000 hive beekeeping enterprise, Miel Carlota in Mexico, so it has strong approval as a tool in large scale commercial beekeeping.

THE HONEY SUPER

The Modified Dadant shallow super has a depth of 6 5/8 inches (168mm), just half way between the depths of the Australian W.S.P. and Ideal supers. The depth of the frames is 6¼ inches (159mm). This is the size that was being used by R.O.B. Manley and for which he modified the original Hoffman frame design to that which has become known as the Manley frame.

What is now known in Australia as the Manley honey super, is a Modified Dadant shallow super reduced in width to that of the Langstroth 10-frame hive, and holding eight Manley frames. It has many advantages which make its use attractive for modern commercial beekeeping.

Beekeepers have recognised that the standard Langstroth 10-frame box is too heavy as a honey super. The strain of lifting these boxes, which average 36kg, has made some beekeepers feel that, at 45, they are too old for honey producing.

The Manley honey super averages 25kg gross, ranging from 23kg to a maximum of 28kg and it is a reasonable and economic weight to lift, carry and load. The average amount of honey in these supers is 18kg, ranging from 16 to 21kg, three-quarters of that held by a Langstroth 10-frame box.

This more manageable weight increases the speed of handling, lessens the fatigue, and eliminates the strain and backache of handling Langstroth boxes full of honey.

Manley super filled with ripe honey

On the hive, the Manley supers are filled and made ready for extracting more quickly. And they are cleared of bees more quickly and thoroughly.

The Manley frames, of simple construction, are especially designed for the storage and easy extraction of honey. With these frames, self spaced at the best spacing for honey storage, there is no 'death rattle' when hives are moved with empty supers. The frames, their end bars

44mm wide, require no fiddling about spacing the frames by hand, and the boxes can be stood on their ends or on their sides without crushing bees or combs.

The width of the top and bottom bars, 28mm, is just right for the removal of the cappings with one clean sweep of the knife. And very little honey is cut off with the cappings. The Manley frames are uncapped in half the time taken to uncap Langstroths, and with a shorter knife and much less fatigue.

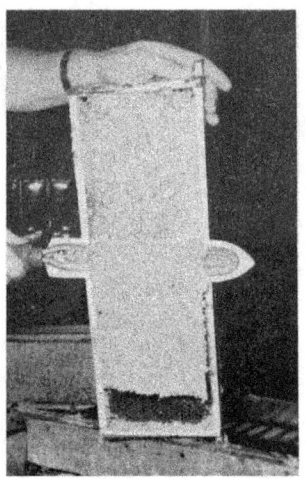

Uncapping a Manley frame

Faster handling and uncapping and the decrease in fatigue and strain, means that more honey can be extracted each day than was possible with the cumbersome Langstroth supers.

Better quality honey is produced. As the combs are never used for brood production, the honey is free from the taint of dark combs, and very little slumgum is produced in the cappings reducer.

Because of the design of the Manley frames, not much honey passes through the cappings reducer. Fuel is saved because less heat is taken up by honey from the knives and from the reducer. Most of the heat is used for melting the wax instead of heating honey. The maximum amount of honey is extracted in the proper place – the honey extractor.

The reduction in slumgum permits a better flow of heat to the wax and results in better quality beeswax.

Finally, the cost of the Manley supers is lower than that of Langstroth boxes. Timber of the width required is more plentiful and cheaper than the wider boards of Langstroth size. The frames are more simple in design, requiring fewer machining operations, and are quicker to assemble.

Those opposed to the use of special honey supers regard interchangeability of combs between the brood nest and supers as being essential. They pay a heavy price for this one feature in denying themselves all the advantages of the use of honey supers designed for the job.

1968 Winter   TRIBUTE TO MANLEY
By Charles Koover, California

If, when I go, I could leave behind a useful piece of bee equipment, like the Manley extracting frame, I would feel that I had been of use to beekeeping as a whole.

I have just extracted my entire crop of honey from Manley frames and what a pleasure it proved to be. Beautiful to behold, they came off the hive unscathed, completely free from burr comb, above and below, and reached the extracting room without a single mishap or leaky comb.

Tightly held in the super; they could not move, yet when it came time to remove them, the super slipped down and forced them out into one block. They separated without the use of force and when the electric uncapping knife slipped under the cappings it could be seen through the thin film of wax as it glided over the comb.

This is beekeeping at its best. What a time-saving invention it is. No honey to be separated from the cappings, so to speak.

Mr Manley should not feel that his life has been wasted. When he goes, he will always be remembered as a practical beekeeper who made a real contribution to the craft. As we say in America: "He left his mark."

*   *   *

Comment by R.O.B. Manley

I have over the years protested that it was not I who invented this appliance, though it has been booked to my credit. As a matter of fact it was when paying a visit to my old friend E.W.D. Madoc in Norfolk that I first saw close-ended frames for extracting and at once realised what a blessing they would be, and changed all ours. But that was many years ago. Honour to whom it is due.

From British Bee Journal 1967

Author's Comment

Mt Manley may not have invented the frame which is called after him, but it was he who brought it to the attention of apiarists in 1946 in his book *Honey Farming*. And this is what he said (pp 67-68) –

"- - - the American type of super frame, considered as a practical appliance to be used in the production of honey, is really an extraordinarily inept idea. It seems strange that these abominations should have apparently satisfied our colleagues across the water for so many years, for although the thing would appear to be sufficiently obvious, it seems never to have occurred to American bee-men that since the super frame is used for an entirely different purpose, it should be designed for that purpose, and not be just another brood-frame.

"I shall be told that these frames are intended to be used for both purposes. They are, over there: which is one reason why they have found it so difficult to control foul brood; but they made their shallow extracting frames, apart from depth, exactly the same as the deep ones.

"The Hoffman frames, so good for the brood chamber, are abominations in supers, for an extracting frame should be spaced wide enough for the combs to be easily uncapped. I have used these frames for extracting by putting nine instead of ten in each [full-width

Dadant] super; but they are troublesome to arrange, and this wastes more time than bee-farmers can spare, and when so arranged you cannot move a super without having the frames move too. So some of us are now using wide, close-ended frames; that is, having their end bars just wide enough so that ten of them fill the super of a Dadant hive allowing a little play." [This is practically the same as having eight frames in an Australian 10-frame super. Ed]

"All our super frames, also, have their top and bottom bars of the same width, 1 1/8 inch [28mm], which is a great assistance in rapid uncapping, the knife passing under the cappings in contact with both top and bottom bars. Why extracting frames are ever made with narrow bottoms, I cannot even guess. Years ago, I widened all mine by nailing strips on each side, since when all new frames have been properly made with wide bottoms, and I certainly never will use narrow-bottomed frames for extracting any more, for the mess caused by brace-comb is generally a great nuisance when the frames in supers have narrow bottoms. Super with such frames as I have described can be handled rapidly without the slightest fear of any movement of the frames, and there are no loose parts or metal fittings of any kind."

*   *   *

Manley's argument impressed me when I read his book soon after it was published. But after I myself had had the experience of working in the apiaries of his 1800-hive bee farm, had removed the full supers using clearer boards with Porter bee-escapes, and had helped to extract 40 tons of honey from these supers in July-August 1949, - I was convinced.   F.G.S.

QUEEN EXCLUDERS

The great benefit of the use of queen excluders is the enormous reduction in the labour of removing honey from the supers. Other benefits are the production of better quality honey because combs are not contaminated by the smell and taste of brood rearing and no combs containing brood go through the extractor. In addition, there is a great reduction in the danger of spreading brood diseases.

Any beekeeper who has used queen excluders properly and experienced the ease of removing boxes full of honey from the hives, knowing that they are free of brood, will never return to the time-wasting labour of picking through individual combs in the super boxes of a hive at extracting time.

Not only does the queen excluder keep the queen below in the brood box, it also helps to keep pollen out of the honey supers. The place for pollen is next to the brood in the brood box, not mixed up with the honey in the supers.

The queen excluder is an essential part of every honey producing hive. It is placed on top of the brood box to keep the queen from going up into the honey supers and laying eggs in them. The excluder is a wire grid in which the holes are of such a size as to permit worker bees to pass through but prevent the passage of the queen and drones. With excluders of the wire type, the space between the wires is specified as 0.168 inch +/- 0.003 inch (4.27mm +/- 0.07mm).

The type of queen excluder recommended is the metal bound welded-wire queen excluder. This is a great improvement on the wooden framed excluder, being stronger and easier to clean.

Some beekeepers find that they obtain more honey when they use queen excluders. Without a queen excluder, the queen is tempted to go up into an empty super and use it for brood rearing, neglecting the bottom box. But with a queen excluder in position, the queen is forced to utilise fully the bottom box for egg laying and the bees are stimulated to fill the super with honey.

A queen excluder can always be put between two boxes in which the bees are already working. But do not put a queen excluder on top of a box which is not fully occupied by bees. If the bees are not ready to occupy a super, then the addition of a queen excluder may delay the use of the super when the bees are ready. If you must give the bees a super before they are ready for it, wait until they have started work in the super before putting the queen excluder under it.

A point to note when using a welded-wire excluder; put the excluder in position with the cross wires uppermost. This ensures that there is only a bee space between the excluder and the top bars of the fames underneath, and that there is sufficient space to stop propolising between the queen excluder wires and the bottoms of the frames in the box above.

HIVE LIDS

The lid or roof, as it is called in the Northern Hemisphere, covers the top of the hive. It protects the bee colony from the weather and from enemies, and is readily removable to give the beekeeper access to the interior of the hive.

The lid must be strong enough to permit hives to be stacked, for a man to stand on top of it, and for the hives to be lashed down on a vehicle. The materials of which the lid is made must be sufficiently durable to withstand the effect of the weather, the full heat of the sun, rain and damp. The lid is required to keep the hive cool in summer and dry at all times. It must stay in position in a gale and it must not provide a nesting place for ants or other pests.

The variety of hive lids in use in Australia was described in the previous chapter. The lid which appealed to me as being the most practical I have ever seen anywhere in the world is the cavity lid, which is in effect a ventilated migratory lid with a built-in inner cover.

The migratory lid consists of a 50mm high rim, with a central cross piece and with boards or tempered hardboard on top, covered with galvanised iron sheet and with ventilation holes in the rim. These are covered on the inside with wire gauze. The rim is made to exactly the same dimensions as the hive body and supers. It rests on top of the hive and, like the hive bodies and supers, it is securely attached to the hive by the bees with propolis and does not blow off in a gale. It provides a ventilated air space over the frames which protects the colony from overheating and severe damp. Being of the same external dimensions as the hive bodies, it simplifies packing hives on a truck.

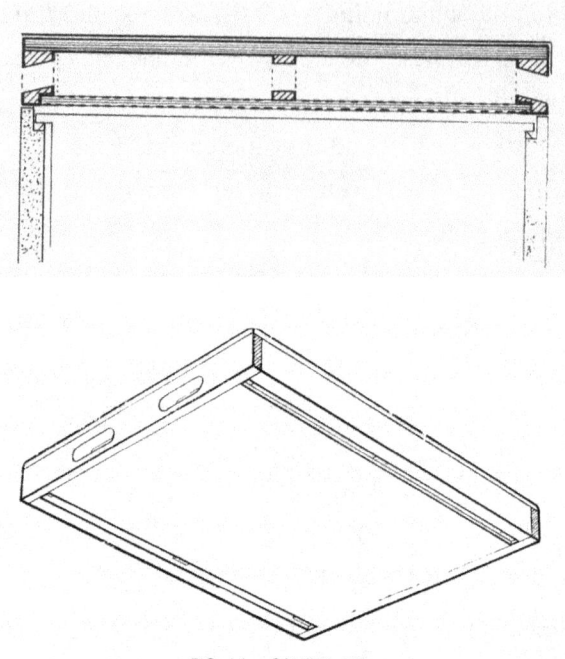

Cavity Lid, ventilated at the ends and with a built-in inner cover with ventilation slots along its sides

With the usual type of migratory lid a mat, consisting of a piece of thick plastic or canvas, cut rather smaller than the inside of the hive, is placed on top of the frames. This is intended to discourage the bees from building comb in the lid. It is stuck down to the top bars of the frames by the bees with propolis so is an additional item to be removed after the lid.

The cavity lid is very similar to the migratory lid in appearance. Instead of having to use a separate mat which sticks to the top bars of the frames, an inner cover is built into the lid, thus halving the time and work required in opening and closing the top of the hive, and with far less disturbance to the bees.

The built-in inner cover consists of a piece of waterproof plywood or tempered hardboards, fitted into and flush with the bottom of the lid. The inner cover is narrower than the inside of the lid to provide a

13mm wide gap along each side for ventilation and to enable the bees to keep the lid cavity free of unwanted guests.

The ventilation holes are provided in the rim at the ends of the lid. In this position the wire screens on the inside of the holes are not blocked up by propolis. Ventilation in the lid is important both in summer and winter. It is the most simple and effective answer to problems caused by high temperatures in summer as well as enabling the escape of damp air, which would otherwise condense on the inside of the lid and on the walls of the hive in cool weather, producing a damp and mouldy hive.

An alternative to galvanised iron is to cover the top of the lid with a hard waterproof plywood. A very serviceable and reasonably priced plywood for this purpose is 8mm thick Karri Aquatite. To prevent splinters, the edges and corners should be rounded by sanding and the whole lid, inside and out, should be painted thoroughly with primer, two layers of undercoat and best quality hard gloss enamel top coat.

BOTTOM BOARDS

During my tour of the Eastern States in 1965 I was most intrigued to observe in Queensland the adoption of baffled entrance bottom boards. Most of the bottom boards had a baffle or false floor, like the built-in inner cover in the migratory lids, with a 13mm space along each side and with bee space above the baffle. K. Mitchell of Warwick and also C.S. Scarfe in South Australia were using the old Miller type of bottom board with the removable frame of slats and a 100mm wide horizontal baffle near the entrance.

My previous experience of the use of this principle was in Dr John Anderson's Glen Hive in the north of Scotland. This had a false floor providing a large entrance at the front of the hive and a slot leading into the interior at the back. In a modification by J.K. Beaton an adjustable roller was fitted at the back which controlled the size of the entrance to the interior of the hive.

The main object of the Glen Hive type of entrance was to provide protection from the weather and to prevent the entry of mice in winter. Dr Miller's slatted floor with a baffle over the entrance was very successful in preventing the wasting of comb space over the entrance. The bees frequently cut away large pieces of comb in the corners of the frames near the entrance when the ordinary type of American or Australian bottom board is used. Dr Miller's floor also provided extra clustering space for the bees under the frames.

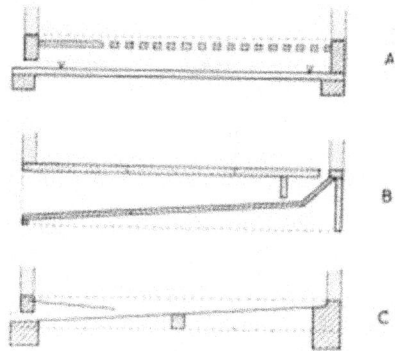

A. *Miller baffled and slatted bottom board*
B. *Protected entrance in the Glen Hive*
C. *Tunnel entrance bottom board*

As I noted in my report on my 1965 tour, the advantages claimed for the baffled floor and the Miller floor are that they prevent the combs from being cut away by the bees near the entrance and they provide clustering space under the combs in hot weather. However, they are also used with hives that have less top ventilation than has been found to be desirable in experiments in Western Australia.

In 1966 I concluded that the baffle or tunnel type of floor offers the following advantages:-

1. The colony is better protected from the effects of winds, both hot and cold.

2. The hive is more readily defended by the bees against enemies.

3. The ventilation of the hive can be controlled by fewer bees.

4. Full use is made of the comb space provided in the frames in the bottom box.

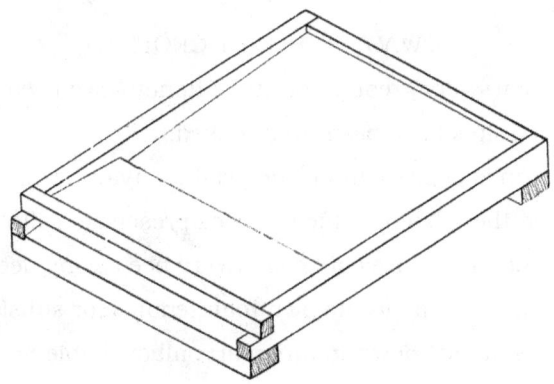

Tunnel entrance bottom board

CHAPTER 17
APIARY SITES

1969   WASTED HONEY CROPS

In recent years very great amounts of honey have been lost because apiary sites have been overstocked.

A honey crop is dependent on the availability of nectar in the flowers and on there being sufficient bees present to collect it. But there can be too many hives of bees. Every bee colony needs a certain amount of food each day for its own maintenance or subsistence. If too many hives are set down in an apiary, all available nectar is used for the maintenance of the colonies and there is none left over to be stored as surplus in the supers.

As many as 100, or 200, or even 450 hives of bees in one apiary may have produced a crop of surplus honey on one occasion, but that will not always happen. Whenever there is a honey flow, it does not mean that 100, or 200, or 450 hives is the right number to collect it.

On many occasions, not only do the bees fail to produce a crop of honey, but the colonies decline in strength because there were too many hives for each to obtain even a bare subsistence. With fewer hives in an apiary, a crop of honey could have been obtained. Instead the beekeeper gets no honey and ends up with weakened bee stocks.

To be successful, an apiarist must produce honey and not be merely a keeper of large numbers of hives of bees. The sustained production of honey depends upon adjusting the number of colonies of bees in an apiary to the nectar which is available. When there is a show of blossom one cannot always put down 100 or 200 hives of bees in the site and get a crop. Such numbers are productive only when there is an exceptional flow of nectar.

Honey flow in the Karri forest

How much surplus honey?

The amount of the honey crop which can be extracted can be expressed as a simple equation:-

AMOUNT OF SURPLUS HONEY = AMOUNT COLLECTED FROM THE BLOSSOMS AND CONVERTED INTO HONEY minus AMOUNT CONSUMED BY THE BEES.

That is –

CROP = TOTAL FROM FLOWERS – FOOD FOR BEES

When an apiarist sets down his hives of bees in an apiary, he does so because there is nectar to be collected from the blossoms. He assumes that the bees will collect it, convert it into honey and store it in his supers. If they fail to do so, it is because of one of two things. Either the colonies were too weak to forage sufficiently and they used what they collected to build up their strength by brood rearing or, if they were at foraging strength, they used all they collected for their own maintenance. Every apiarist knows that he needs to build up his bees **for** a honey flow and not **on** a honey flow, and he will not expect weak colonies to produce a surplus for the first few weeks of a flow.

The situation we are examining is that in which an apiarist sets down strong colonies within reach of nectar yielding blossom, and finds that he gets little or no crop and that the colonies may also decline in strength.

I estimate that an average colony of bees in the active season, maintaining its strength with six or seven frames of brood, consumes an average of 700g honey each day. It could be 900g or it could b e 450g, but that would not affect the principle of my argument. Nevertheless, I believe that 700g per day is pretty close to the mark.

Production per day

Apiarists say that during a good honey flow the hives fill up and are ready for extracting in ten days. What precisely is "full up"? How many supers are on the hive? This varies between beekeepers, between apiaries and between hives. But a figure frequently used is one tin per hive, or 27kg of honey. If an average of 27kg per hive could be extracted after ten days, that would be considered a very good honey flow.

This, then, is equivalent to the bees storing surplus honey at an average rate of 2.7kg per day per hive. I have actually observed a maximum net gain of 5.4kg in a day. I believe that 6.8kg has been recorded on occasions. But these figures are exceptional and occur only when the peak of foraging strength of a colony coincides with the peak of nectar secretion.

More commonly, extraction of honey after ten days yields half a tin, or 14kg per hive. This indicates that the storage of honey was at the average rate of 1.4kg per day.

I would suggest that this last is the more normal figure and that the higher figure of 2.7kg per day is achieved only under exceptional conditions of nectar secretion and when colonies are in first-class condition after a good build-up.

Total available honey

If from an apiary of 100 hives, 27kg of honey per hive can be extracted after ten days, then the total gathered by the bees, in terms of nectar converted into honey per day, is as follows:-

Surplus at 2.7kg per day per hive	270kg
Subsistence at 700g per day per hive	70kg
Total honey gathered per day	340kg

If however, only 14kg per hive is extractable after ten days, the total gathered is as follows:-

Surplus at 1.4kg per day per hive	140kg
Subsistence at 700g per day per hive	70kg
Total gathered per day	210kg

These are good honey flow conditions.

Having obtained some idea of the quantities of nectar converted into honey which can be gathered from a single apiary site under good and very good honey flow conditions, we will now consider the relationship between numbers of hives in an apiary and the extracted honey crop.

Numbers of hives and honey crop

Under very good honey flow conditions we saw that the total honey gathered by the bees was at 340kg per day and that yielded a surplus of an average of 2.7kg per day per hive, or 27kg per hive at the extraction after ten days, with 100 hives on the site.

If 340kg per day was all that was available to be gathered, we will see what happens when we put down another 100 hives alongside the first load.

Total honey available per day	340kg
Subsistence at 700g per hive/day for 200 hives	140kg
Total available to be stored as surplus	200kg

Divided among 200 hives = 1kg per hive per day

By raising the number of hives in the apiary from 100 to 200, the total surplus honey being stored has been reduced from 270kg per day to 200kg per day, and the production per hive reduced from 2.7kg to 1kg per day.

It would be better to place the additional 100 hives on another apiary site where they, too, could store a surplus of 2.7kg per hive per day, and so raise the total production from the 200 hives to 540kg per day, instead of reducing it to 200kg.

It is common to find an apiary of about 100 hives in which the bees are working steadily, maintaining their brood nests, but storing no surplus. The apiarist blames the trees, the weather, the strain of bees and anything else. I would suggest that such an apiarist is merely a keeper of bees, not a producer of honey.

Let us look at honey availability. If there is enough coming in to enable the bees to maintain their brood nests, but store no surplus, then –

Subsistence for 100 hives of bees at 700g per day = 70kg

The bees are collecting in that apiary area the equivalent of 70kg of honey per day. But the beekeeper is getting nothing.

If the apiarist reduces the number of hives of bees to 50, what will happen?

Subsistence for 50 hives at 700g per day = 35 kg

Honey available for storing as surplus   = 35 kg

or 700g per hive per day = 7kg per hive in ten days.

By putting the other 50 hives on another similar apiary site, they could also be collecting 35kg per day, making a total of 70kg per day being stored by the 100 hives in two separate apiary sites, instead of nothing when all together in one apiary.

We could take this further, and split the load up into four separate apiaries of 25 hives in each.

In each apiary, assuming the nectar secretion situation is the same, availability at 70kg per day:-

Subsistence for 25 hives at 700g per day = 17.5kg

Honey available for storing as surplus   = 25.5kg

or 2.1kg per hive per day = 21kg per hive in ten days.

This would be considered a pretty good honey flow! If each of the four apiaries of 25 hives did the same, there would by a total of 210kg of honey a day being stored by the 100 hives instead of **nothing** if they were all in one apiary.

How many beekeepers have had to give up because they would insist on putting down a truck load of hives in each apiary site.

Conclusions

I am not suggesting that all apiaries everywhere and at all times should each contain only 25 hives. There are occasions when there is enough nectar available for 100 or even 200 hives to store their maximum surplus. But this is where the apiarist's experience as a honey producer comes in. For he must be able to judge from his own past experience of the flora in each apiary area just what the optimum number of hives of bees is for each apiary at any one time.

There have been those who, on being advised about apiary stocking, complain that it is difficult to split a load of bees (one of the disadvantages of travelling with open entrances), or that it makes more work when it comes to do the extracting!

What is it to be: make a living from honey production or avoiding work and producing no honey?

1969   FROM A LETTER TO THE EDITOR

Thank you for your *Apiculture* which comes regularly to hand and is much enjoyed for its most useful matter to the industry. That article on 'Wasted Honey Crops' is long overdue and if put into more general practice should make for better returns in production and fewer disappointments to the producer.

H. GRAHAM SMITH,
Former Apiculturist,
Dept. of Agriculture.
New South Wales.

CHAPTER 18

BEE BREEDING

Over the years I had collected a lot of thoughts on this subject; I think that I learned much from Brother Adam of Buckfast in Devon as well as from R.O.B. Manley.

Some breeders, and even commercial beekeepers, in my view had lost their way, particularly those who were concentrating on appearance as the main selection factor; "producing pretty bees" I called it.

So when something triggered my perception of the need, I wrote the following article for *Apiculture*.

1969   SELECTING BEES FOR BREEDING
             THE IMPROVEMENT OF THE BEES

The improvement of bees by breeding should be given top priority by all beekeepers who wish to be progressive. The foundation of successful beekeeping is the bee itself; it is impossible to obtain an increase in honey production without bees capable of the best results. The improvement of the bees by means of breeding is the essential condition for the best results in honey production. Each and every beekeeper can, and indeed must do it to reach and maintain a level of economic production. It is not difficult; it does not require complicated equipment; it merely requires a simple understanding of the principles involved.

Hereditary and environmental factors

Bees in the wild vary enormously in their characteristics. This is Nature's way of ensuring the survival and spread of the species. By numerous crossings between strains, an abundance of hereditary factors is produced which provides for whatever eventualities may occur. Differences are to be seen in all the various characteristics of bees. The differences occur in the form, size, shape and colour, and in the function, development and behaviour, as well as in the ability to store honey. These differences always occur in bees which breed in the wild state and they are hereditary. The indigenous bees in an area are not necessarily the ideal bees for the honey producer, because performance and productivity are not the main objects of nature. In all races, the bad strains, from the beekeeper's point of view, are more numerous than the good ones.

Environment has its effect of behaviour and performance and on the selection of those strains which will survive and those which will die out. But the differences in production in the same apiary during the same honey flow, and other differences in form and function when environmental factors are equal, are due to hereditary characteristics. The honey bee is no exception to the general laws of genetics formulated by Mendel.

Bees which are so ready to defend their hives that they are uncontrollable when kept in hives in the open may be perfectly manageable when the hives are in a bee-house. Here the environment of the cool and shade and the inability of the guard bees to get at the beekeeper to start stinging make the bees manageable. But it does not remove the hereditary factor of bad temper.

However, if the beekeeper steps in and removes the queens in the worst-tempered colonies and replaces them with queens raised from the best-tempered colonies, and if he continues to do this, the temper of his bees will improve. Here the beekeeper by his breeding is introducing a factor into his apiaries which can be called environmental. The strains in which the hereditary factor of bad temper is most marked are eliminated and only the good-tempered colonies and their progeny are allowed to survive.

Similarly, if the queens of those colonies in an apiary which produce the poorest crops are removed and replaced by progeny from the queens having good records of production, and this process is continued all the time, the average production per colony will be increased. Some of the progeny of queens and drones with a good history of production will lack the hereditary factors of fecundity and industry, while other will have one or the other, or both, in accordance with the general laws of heredity. The beekeeper then breeds only from those which have the desirable factors well developed, and eliminates the others.

The aims of breeding

The final aim of breeding is to a certain extent determined by climate and the conditions of honey flows and by the particular needs of individual beekeepers. But there are certain fundamental principles which are universally valid and independent of local conditions and requirements.

The final aim of bee breeding is the production of bees which give us a **constant maximum average yield of honey per colony with the minimum expenditure of time and money.** To attain this aim, there are **four** vital characteristics which must be regarded as the primary objectives in bee breeding, namely:-

1. Industry in foraging. The first priority must be given to an inexhaustible capacity for work. Bees are variable in this respect, and capacity for work is a hereditary factor and one which depends on many other hereditary factors.

2. Fecundity. Adequate fecundity is essential, since the best crops can be gathered only with colonies of maximum strength. Fecundity of itself is not a decisive factor, but it is an indispensable basis for first-class performance. We do not want bees which turn every pound of honey into brood or which breed excessively during periods of dearth, and there are such bees. We require bees which will build up their populations rapidly for the honey flow, and which will maintain the colony strength during the flow. We do not want bees which only build up on the flow.

Queens, which cannot fill ten or eleven Langstroth combs (or eight or nine full depth Dadant or Jumbo combs) with brood during the build-up period, do not have the qualifications necessary for our purpose.

3. Resistance to disease. One of the most important tasks of breeding is the development of strains which are resistant to disease. The advantages of breeding resistant strains is clearly seen in the case of Acarine disease in Europe. Resistance to American foul brood is closely linked to a highly developed instinct for cleanliness in the hive. Resistance to disease saves the work and cost of applying remedial measures. The use of anti-biotics and other such remedies may check the disease but their effect is not lasting. Further, the use of such remedies conceals which bees have a resistance to disease.

4. Disinclination to warm. The last of the indispensable qualities is the disinclination to swarm. From the commercial beekeeper's point of view this is essential. Swarming not only causes a great deal of work and a loss of time; it may also ruin the chance of obtaining a honey crop. A race or strain of bees which possesses all the other desirable qualities and yet is prone to swarming is useless for modern beekeeping because all the other good qualities are wasted.

Secondary objectives

Industry, fecundity, resistance to disease and disinclination to swarm should form the basis of our breeding. There are other characteristics which, although not so essential, are of importance as each helps towards greater production.

1. Long life. There are hereditary differences in the life span of queens and of workers. Apart from heredity, the treatment given by the nurse bees, especially during the feeding period, has an influence on the length of life of both queen and workers. There tends to be a relationship between fecundity and longevity. The over-prolific races tend to be short lived, and great longevity is found among the least prolific bees. Great longevity combined with lack of fecundity is a disadvantage when parasitic infestation shortens the life of the bees, and may lead to extinction of the race. The need is to produce a balance of adequate fecundity with good longevity.

2. Flight range. This has an important bearing on the foraging ability of the bees. The longer the distance they can fly the larger the area of country they can cover in search of nectar sources. Flight range or wing power can determine whether or not a crop will be produced from a distant source of nectar.

3. Keen sense of smell. This also improves foraging ability and enables bees to find sources of nectar which are missed by bees lacking keenness in sense of smell. But this characteristic easily leads to bees robbing from other hives. The best nectar gatherers are normally the first to start robbing.

4. Instinct for defense. The most reliable remedy against robbing is good instinct for the defense of the colony. This is essential against intruders such as ants, wasps, wax moths and other enemies. It is most highly developed in the honeybees of Africa and the Middle East, and to such an extent that it makes management difficult if not impossible. The instinct for defense must be balanced by good temper.

5. Hardiness. Bees must be able to withstand the cold and damp of the winter months as well as the high temperatures of summer. They must be quiescent in unfavourable weather when to fly out would lead to loss of life. This is associated with other qualities such as the ability to keep the hive dry in winter, which varies considerably from one colony to the next, and freedom from restlessness caused by sharp changes in temperature and disturbances.

6. Conservation of stores. The consumption of stores in winter and during other periods of dearth is bound up with quiescence in winter, the strength of the colonies and fecundity. Bees which are extravagant in the use of their stores during periods of dearth are frequently checked during the build-up because they have exhausted their reserves of food, and consequently build up on the honey flow.

Economy in the use of stores during periods of dearth is essential.

7. Spring build-up. The nature of this characteristic will vary according to the district. In general, build-up after a period of dearth should commence as soon as pollen becomes available, provided that the weather is suitable, and then go forward to reach a peak at the beginning of the honey flow.

8. Pollen collection. Races of bees vary considerably in the quantities of pollen they gather. Some gather pollen greatly in excess of their needs, filling up the brood combs and even taking it up into the supers through queen excluders. Pollen should be collected in quantities adequate for their needs and stored immediately above and to the sides of the brood.

9. Comb building. This is a characteristic which has a direct influence on honey production. Colonies which are slow to build comb are inclined to swarm. There must be a keenness to build worker comb, for the expansion of the brood nest and for the storage of honey. But the building of excessive amount of drone comb is wasteful.

10. Arrangement of stores. This is closely bound up with the building of comb. Bees reluctant to build comb store the honey close to the brood nest. The characteristic required is an inclination to store honey away from the brood nest during the honey flow periods, coupled with the instinct to store in the brood chamber at the end of the flow and before winter or other periods of dearth.

11. Quality of stores. Some races and strains are inclined to collect better quality honey than others, from the same flow. This should also be kept in mind when breeding.

Management characteristics

The characteristics discussed so far have a direct bearing on honey production. There are other characteristics which do not influence production, but are essential for economy of management, the minimum expenditure of time and money.

1. Good temper. Bad-tempered bees make the work harder, cause loss of time and may produce unpleasant incidents. Good temper is a hereditary factor which can be bred into a strain.

2. Calm behaviour. Calmness and steadiness of the bees on the comb are a great help for easy working. The bees and queen should remain fairly still on the combs and not rush round when under manipulation.

3. Disinclination to propolised. The gumming-up of the movable parts of the hive with propolis considerably increases the work of the beekeeper, but it is a difficult trait to eradicate. A reasonable application of propolis is necessary for those who use migratory lids; it stops them being blown off in a gale.

4. Freedom from brace comb. Like propolis, brace comb construction makes the manipulation of frame hives difficult. Breaking the brace comb between boxes and between combs often requires a great deal of force and causes bees to be crushed and may make the bees bad tempered. It is a trait which can be removed by breeding.

5. Disinclination to drift. This characteristic helps to enable a bee to find her own hive. It is important for beekeepers who set down large numbers of hives close together. Lack of this characteristic causes heavy losses of queens on mating flights, and unevenness in hives' populations; hives at the ends of rows gaining bees at the expense of hives in the centre.

6. Cleanliness. A well-developed instinct for cleanliness in the hive is a great help to the beekeeper. It is most important in the prevention and control of brood diseases and in resistance to wax moth. Bees which tolerate half-mouldy combs have a very low instinct for cleanliness.

7. Good cappings. Even and level cappings on honey comb are a great help to the beekeeper when extracting. White cappings, which have an air space between the honey in the cell and the cappings, are essential for those who market honey in the comb. White cappings are also a help to producers of extracted honey as there is less honey to be separated from the wax after the cappings have been cut off.

These are the qualities which have to be taken into consideration in breeding. External characteristics of the bees such as colour are pointers to a right assessment of the purity of a strain, but they must never be regarded as infallible guides to performance.

Performance is not due to any one factor alone, but to the interaction of a whole chain of factors. The more perfect the interplay of the individual factors in the chain, the higher the standard of performance.

Stock from overseas

It is necessary for us to be on our guard against certain practices carried out by some queen-breeders overseas. These are the wholesale feeding of antibiotics to bees and selection for fecundity or colour without giving sufficient consideration to the other necessary factors.

The wholesale feeding of antibiotics can lead to the propagation of strains of bees which are lacking in resistance to diseases. Such bees may be highly susceptible to infection by disease organisms when deprived of the antibiotic protection. Health certificates issued in respect of such bees are virtually worthless because the presence of viable disease organisms may be masked by the drugs.

While on the subject of diseases, it is necessary to mention Acarine disease. The bees in Australia could be highly susceptible to Acarine disease as are the bees in America. In western Europe, bees have developed a very high level of resistance to Acarine. It is essential that precautions continue to be taken against the importation of Acarine disease into Australia. With regard to external mites, these are present in Australia and in North America, and they have not caused Acarine disease to develop in either of those countries. Further, there is no evidence that they have caused Acarine disease in Europe. All parasites are undesirable, but we should not be unduly worried about the external mites. However recent research suggests that they may in some circumstances carry organisms which can cause colony collapse.

Bees bred mainly for fecundity or for colour, without consideration of productivity and the other factors, are worthless. We do not want our hives boiling over with bees that eat up our honey crop. We want bees which will fill our hives with honey.

Acclimatization of imported bees

Not all bees imported into Australia, nor that that matter transferred from one part of Australia to another, will be of value, even if they do come from the best stock. Such bees have to adapt themselves to new conditions, and only a small percentage will be able to make the required adaptations.

For one thing, bees brought from the northern hemisphere to the southern hemisphere have to learn that the apparent movement of the sun is from right to left instead of from left to right. If they do not, they will be lost when returning from foraging. They must also learn to adapt themselves for new climatic and honey flow conditions.

So, when new stock is imported, or transported to new environments, selection of the adaptable strains is essential, and breeding must be continued in the environment in which the bees will be used for production.

CHAPTER 19
BEESWAX

Earlier, when I was doing my first inspections of honey extracting plants in Western Australia, I commented on the beeswax produced as follows:-

"Then there was the problem of the beeswax. Why was so much of the wax produced by commercial beekeepers a dark greeny-grey colour and resistant to bleaching? It was nothing to do with the plants or the pollen which gives beeswax its natural colour; that was demonstrated by Rob Smith, a leading beekeeper who consistently produced beautiful cakes of yellow beeswax for the honey shows. The dark stained beeswax reminded me of the beeswax produced in Northern Rhodesia, which had been melted by the beekeepers in iron cooking pots, and by the traders in iron drums."

1964   ADULTERATION OF COMB FOUNDATION

When any other type of wax, oil or fat is mixed with beeswax, the beeswax is adulterated and it is impossible to separate the pure beeswax from the adulterant. Adulterated beeswax is fit only for certain low grade uses for which almost any wax would do, so it is of little value. The cosmetic, pharmaceutical and industrial uses which require pure beeswax and pay good prices for the pure unspoilt product, are completely closed to adulterated beeswax.

There is a temptation for comb foundation manufacturers to add cheap petroleum wax to their beeswax. This happened in America some years ago and it threatened to undermine the market for pure beeswax. Eventually all the American manufacturers signed an agreement to use only pure beeswax in comb foundation.

Beekeepers are confronted with enough problems over their wax without adding adulteration. All are urged to demand from their suppliers of comb foundation a guarantee that it is made entirely from pure beeswax.

1969   DAR ES SALAAM BEESWAX

Why does Dar es Salaam beeswax fetch a better price on the London market than good Australian beeswax? This question has been put to me several times in recent months. Having devoted thirteen years of my life to the improvement of Dar es Salaam beeswax I am in a position to answer this question.

Beeswax produced in Tanganyika and exported through the port of Dar es Salaam is known as Dar es Salaam beeswax. It has a long established reputation for good quality. This was started by the Germans at the beginning of the century when they taught the African beekeepers how to render beeswax. At that time the Germans were leaders in the wax refining field and the principal British wax refiners learned their trade from the Germans. The standards established by the Germans were maintained by the business houses dealing with the export of beeswax from Dar es Salaam. Later the Tanganyika Government played their part by doing extension work on beeswax preparation and by establishing export standards for beeswax quality. This work was intensified and expanded in 1949 when I arrived in Tanganyika to study African bees and beekeeping methods and to improve and stimulate beeswax production throughout the Territory.

The main reason for the popularity of Dar es Salaam beeswax is that it bleaches easily. This is because the colouring matter in the wax is only that which it has obtained naturally from the oils in pollen grains. It is free from the unbleachable stain caused by chemical reaction between beeswax and certain metals. It is also free from the damage caused by overheating and by prolonged heating.

Overseas buyers have always been insistent that producers should be taught to render beeswax into clean cakes in the first place, so that no subsequent treatment is required. Dar es Salaam beeswax is exported in the form of the original cakes as rendered by the beekeeper, broken in half or perhaps smaller pieces by the shipper to ensure that there are no stones or other foreign bodies in the cake. The overseas buyers have been most insistent that cakes of wax of different colours should not be melted down together because the refiners like to be able to pick out different colours for different uses. The pale yellow waxes are used for making cosmetics and for pharmaceutical purposes, the orange and darker colours go into polishes, electrical insulation and for other purposes where colour does not matter. The overseas buyers have always preferred to tolerate a little bit of dirt rather than have large quantities of wax melted down in iron or galvanised iron drums and subjected to prolonged heating.

Tanganyika has been fortunate in that the domestic appliances readily available for the beekeeper were made of materials which do not react chemically with beeswax. Earthenware cooking pots, good quality tin-plated four-gallon kerosene tins, aluminum cooking pots and enamel basins are the vessels normally used for rendering beeswax by the Tanganyika beekeeper. None of these hurt beeswax in any way. The wax is melted in plenty of water but the water is never allowed to come to the boil because this would cause a partial emulsification of the wax. The wax is constantly stirred in the water while being heated and as soon as all the wax has melted, the fire is withdrawn.

Another important point is that the bowls used as moulds for the melted wax, after it has been strained, are smeared with a film of soapy water or, in some cases, with honey. Honey does not damage the wax but has the disadvantage of leaving the cakes sticky and attractive to foraging bees. Oil or fat is never used for this purpose. It must be remembered that while ordinary solids like sand and dust can be removed from beeswax by filtering, wax that has been damaged chemically by reaction with metal or by overheating, or wax which has been adulterated by the addition of another wax or an oil or a fat cannot be made good by refining. It is permanently damaged and its usefulness is greatly limited, or lost altogether.

There is no reason whatsoever why the wax obtained from cappings in Australia should not be equally good and in as much demand and fetching as good a price as Dar es Salaam wax. All that is required are capping reducers and moulds made of metals which are not injurious to the wax. In fact I have seen wax in Western Australia which has been in contact only with stainless steel and is quite comparable with Dar es Salaam wax.

The wax from old brood combs presents altogether a different problem. It should never be mixed with the wax from cappings. The first problem lies in the comb foundation. This was inserted by the beekeeper to provide a base on which the bees built their comb. But in most cases the beekeeper has no knowledge of the composition of the wax or of the treatment to which the beeswax forming the comb foundation had been subjected before he bought it. It could well be adulterated although that is unlikely. It probably has been overheated, and it most certainly will at some stage or other have been in contact with galvanised iron or other injurious metal.

Then there is the problem of the wires in the frames. These wires are poorly coated with tin or are galvanised. The steel is frequently exposed. The nails used to hold the frames together may have been cement coated but by the time the old comb is ready for rendering the nails have become rusty. So it can be seen that when the old combs are melted there will be a reaction between the wax and the frame wires and the nails. This detracts from its quality.

It is difficult to see how any improvement can be made in this field until-

(a) manufacturers of comb foundation use only cappings wax for the manufacture of foundation;

(b) frames and foundation have wire which is very well tinned or is of stainless steel or monel; and

(c) the parts of the frame are secured by nails which do not corrode, or by a type of glue which is not soluble in hot water.

Until such time as this occurs the wax from old combs will have to be kept quite separate from the wax from cappings. This will enable Australia to market cappings wax of a quality comparable with Dar es Salaam wax and to use its wax from old combs for other manufacturing purposes which don't require such high standards as the cosmetic and pharmaceutical industries.

The main points to remember when rendering beeswax are as follows:-

(a) Iron, zinc, galvanised iron, brass or copper should not be used for the vessels or for the metal parts coming into contact with melted beeswax because they tend to stain it.

(b) The wax should be heated either in water or with a water jacket between it and the source of heat. Overheating causes decomposition of the wax and prolonged heating darkens its colour.

(c) Old combs and dark combs should be melted in water to improve the colour.

(d) Cappings and wild comb such a burr comb and brace comb, should be melted separately from the old combs to prevent the wax of the cappings and wild comb from being stained by the combs which are dark or already damaged. Beeswax of different colours should not be melted together.

(e) Soapy water may be used for smearing in the moulds to prevent the wax from sticking when it sets, but under no circumstances should any oil or fat be used for that purpose. If the wax is not too hot and the surface of the mould is quite smooth, then there may not be any need to use even soapy water.

If these simple precautions are followed there is no reason why Australia should not be able to market its wax from cappings and from wild comb on terms as good as those obtained for Dar es Salaam beeswax.

Briefly:

1. Use vessels which do not react with melted beeswax.

2. Avoid overheating or prolonged heating of the wax.
3. Melt cappings and white comb separately from old dark combs.
4. Melt old combs and dark combs in water to improve the colour.
5. Avoid contaminating the wax with any oil, fat or another wax.
6. Use salt-free water (rain water) for rendering wax.

1970 Autumn   BEESWAX PRODUCTION

In Australia we are in a fortunate position of being able to produce our maximum output of beeswax at the time of the year when very little is available from the main exporting countries. From February to June each year beeswax tends to be in short supply on the world market. The latest quotation for Dar es Salaam beeswax dated February 21 (1970), is at £785 sterling per ton c.i.f. Offers are reported to be very scarce as stocks at origin are negligible.

Good quality beeswax, that is beeswax with its bleaching qualities unimpaired, is always in good demand. As our seasons permit us to produce the maximum amount of beeswax at the time of the year when prices are at their best, it is worth while making an effort to produce and market the maximum possible amount of beeswax.

Recently a major overseas refiner has asked if it is possible to get Australian beekeepers to produce beeswax in a similar condition to that obtained from Tanzania (Dar es Salaam), in mixed colours and in smaller pieces. The over-rendering by Australian beekeepers only spoils the bleaching qualities. Particular mention was also made of the damage done to beeswax by the addition of acid and by boiling causing a partial emulsification of the wax and adding to its water content.

With the price of honey, especially the darker grades, being so low, it might well be profitable to feed back medium and darker grades of honey to get the bees to convert it into beeswax.

CHAPTER 20
RESEARCH IN WESTERN AUSTRALIA

POLLEN ANALYSIS

Soon after arriving in Western Australia, I set up a laboratory and began studying the pollen grains of the principle honey producing plants. I used the same methods as I had used in Tanganyika and carried out pollen analyses of honey samples from all the beekeeping areas of WA.

Some work had been done earlier at the University of WA on pollen grains in relation to ancient deposits. The author of that work had found that he could not find sufficient distinction between the different species of Eucalyptus for his purpose.

I found that the main forest species of Eucalyptus were easy enough to distinguish and with certainty when the location was known, but there were problems with some species which occurred in the wheatbelt and eastern goldfields and were closely related to each other.

At the Open Day at the Department of Agriculture the bee forage studies were illustrated with a collection of Eucalyptus from the Goldfields and east of Hyden, together with Dr Beard's vegetation maps, the descriptions and photographs of pollen grains and the identification of pollen grains in honey and pollen loads.

I demonstrated the preparation of pollen grains for microscopic examination together with honey analysis procedures, particularly in respect of pollen analysis, colour grading, determination of water content, and reducing sugar and sucrose analysis.

The collection of pollen grains from living plants enabled me to observe and to photograph the division of the pollen grains of *Dryandra nivea* which have only two apertures, a process not previously recorded.

SMITH, F.G. (1968) "Dryads" in the Proteaceae *Grana Palynologica* 8 (1): 86-87

1965   HIVES IN HOT CLIMATES

A series of experiments was carried out during the summer 1963-64 at South Perth where the air temperature in the shade in summer often exceeds 32°C and is commonly over 38°C. Tests were made to determine the effect of certain treatments on the temperature within the hive. Among the treatments tested were painting hives with different colours, both internally as well as externally, the use of inner covers and different types of hive lids with various amounts of top ventilation. Some investigation was also made on the effect of different sizes of bottom entrances.

The essential points arising from this work are as follows:-

1. The environment of an apiary on bare ground on a sunny summer's day is about 5°C higher than the official air temperature in the shade.

2. The inside of a hive painted with bituminous-based aluminum paint is up to 5°C hotter than a white painted hive, and one painted with oil based aluminum paint is up to 3.6°C hotter than a white hive.

3. In hives with ventilated lids, aluminum paint on the boxes of two deck hives makes them up to 4°C hotter inside than if they are painted white. Aluminum paint on a ventilated lid, or on the outside of a bottom board, adds about 0.6°C to the temperature in each case.

4. Painting hives internally makes no appreciative difference to the maximum temperature in the hive.

5. In hives without lid ventilation, the use of an inner cover makes no apparent difference to the maximum temperature in the hive.

6. The coolest type of lid is one which is painted white and has top ventilation permitting air to flow from within the hive to the exterior.

7. The lowest maximum temperatures occur in white hives with 58sq cm (four vents each 75mm x 19mm) or more top ventilation.

8. In aluminum painted hives, increasing the top ventilation

reduces the maximum temperature within the hive, but even with 116sq cm (four vents 150mm x 19mm) top ventilation, aluminum painted hives are 1.67°C hotter than white hives with the minimum of 15.5sq cm ventilation (four 22mm diameter holes).

9. Increasing the size of the bottom board entrance to substantially larger than is normally used in Western Australia increases the maximum temperature within the hive. The coolest hives are those with the smaller entrances to restrict the inflow of hot air, but with the most top ventilation to let hot air out.

All these experiments were done out in the open under natural apiary conditions. In order to measure the full effect of the different treatments on the environment within the hive, the hives were tested empty, without combs or bees. Combs, especially combs full of honey, have a buffering effect on the temperature within the hive, being slow to heat and slow to cool. Also the presence of bees, by their fanning and by evaporation of water in the hive, exert a substantial amount of control on the temperature.

Nevertheless it is the beekeeper's interest to have the minimum number of bees employed on temperature control, and the maximum engaged in producing honey.

Further investigations

The work described above showed that the provision of top ventilation, from the interior to the exterior of a hive, makes the interior of the hive cooler than an unventilated hive, even if that hive be painted white. The opportunity was taken during the hot spell at the end of 1964 (six days over 38°C), to determine whether the provision of top ventilation had any advantage over merely shading the top of a hive without providing top ventilation. The hives were painted white.

The experiment showed conclusively that the maximum temperature in the centre of the upper box was hotter (by up to 1.1°C) in the hive with a lid which merely shaded the unventilated inner cover. The temperature on the bottom board in the unventilated hive was up to 0.9°C hotter than in the ventilated hive. Identical materials were used in the construction of the hives and their lids in both treatments.

In the hive with through top ventilation (64.5sq cm), the maximum temperatures on top of the ventilated inner cover, in the centre of the upper box, and on the bottom board, were the same, and these were the same as the shade temperatures in the apiary.

The results of these experiments indicate that in hot climatic conditions, where high temperatures can kill the bees or reduce their productivity, and the hives have no protection from the direct rays of the sun, it is advisable to paint all exterior surfaces of the hives white, to provide at least 58sq cm of through ventilation in the lids, and to avoid larger than normal entrances in the bottom boards. If the apiary, including the ground around the hives, is shaded from the sun, the colour of the hives is less important.

Reference:

SMITH, F.G. (1964) The Hive Environment in Hot Climates *Journal of Apicultural Research* 3(2): 177-122

1966   HIVE VENTILATION IN WINTER

An experiment was set up in the departmental apiary at Yanchep to determine the effect of providing through ventilation in the lids of hives in winter, compared with hives with insulated lids and having no top ventilation.

Periodic examinations of the hives were carried out to record differences in amounts of condensation, colony development and the weight of the hives.

There was no evidence of any difference between the ventilated and unventilated groups in rate of development or in production in the subsequent honey flow. The total net increase in weight of each of the two groups was exactly the same.

The hives which had insulated lids without top ventilation

contained very much more condensation. Fungus developed on the top bars of the frames, and paint had lifted on the outside of supers.

The hives which had fully ventilated lids had a little condensation during bad weather, but no fungal growth or damage to paint.

The experiment confirmed that 64.5sq cm of through ventilation (four slots approximately 19mm x 75mm) in the lid of a hive had no adverse effect on colony development or on production during the winter and spring, and that it protected the hive from damp and fungal damage.

Later a small trial was carried out at Dwellingup in Jarrah forest during a winter of particularly heavy rainfall. Other than confirming my earlier finding concerning ventilation, it was clear that the Dwellingup Jarrah forest was no place to winter bees when the valuable flora of the coastal plain was accessible to a beekeeper.

1965 ECONOMICS OF BEEKEEPING

Beekeepers who are interested in going in for honey production for a living need to know how many hives they need to run, the cost of the hives and the necessary equipment, and what return they can expect for their investment and labour.

Variable factors to be taken into account are:-
1. The price at which hives of bees can be obtained
2. The distances (and therefore the expenses) involved in moving apiaries from honey flow to honey flow,
3. The amount of honey harvested,
4. And the price at which the honey can be sold.

If all the factors are favourable a good return may be obtainable, but it needs only one aspect to be unfavourable for the situation to be reversed.

The presentation of costings can only specify theoretical cases based on prevailing conditions in the area under consideration. The main value of these costings is that they show the reader what he has to include when working out his own position.

I received figures and most helpful suggestions from a number of leading beekeepers in Western Australia. Examples of beekeeping finances were given for beekeepers owning 20 hives, 200 hives and 500 hives. Costings and returns were those prevailing in 1965.
Reference:

F.G. Smith (1965) *Economics of Beekeeping* Perth: Dep. Agr. Bull. 3354

1966   SUCROSE CONTENT OF HONEY

Summary

An abnormally high sucrose content was found in some samples of honey produced in Western Australia during the winter of 1964. To determine the origin and possibly the cause of the abnormality, samples from combs and storage drums were analysed. During the following winter, samples were collected from hives, periodically.

The cause of the high sucrose content was found to be a natural one and not malpractice nor bad beekeeping. The high-sucrose honey originated from nectar of *Banksia menziesii,* with *Banksia sphaerocarpa* an additional source in limited areas. Ripe honey from the excellent flow from the former source in 1964 contained 8 to 12 per cent sucrose; one sample of *B. Sphaerocarpa* honey had 20.5 per cent. There was only a minor flow from the Banksia in 1965 and the sucrose content of the honey was between 4 and 9 per cent.

The sucrose content of some, but not all, unheated honey samples which had been held in store for a year decreased to about half its original level through the action of enzymes present in the honey which converted the sucrose into dextrose and laevulose.

No other major source of nectar known to us in Western Australia results in honey with a sucrose content above the normal upper limit of 2 per cent.

It must be concluded that specifications of honey which define it as containing not more than 5 per cent sucrose would therefore exclude some types of true honey, which are rarely encountered in commercial practice, but which are nevertheless pure unadulterated honey. Commercial crops of *Banksia menziesii* honey may contain 10 per cent sucrose. Packers wishing to sell such honey on markets which specify a lower sucrose content can blend the high-sucrose honey with honey of a very low sucrose content, or hold the crop in store of about a year from the time it was produced. Both methods cost money and may be difficult to arrange.

Specifications defining the sucrose as well as the dextrose and laevulose contents of honey should not be too rigid and should make provision for variation, so as to permit trading in pure high quality honey from uncommon sources.

Reference:

SMITH, F.G. (1965) The sucrose content of Western Australian honey *Journal of Apicultural Research* 4(3): 177-184

1966 THE COST OF PROTEIN IN POLLEN SUBSTITUTES

Beekeepers had been confronted by conflicting claims by the proponents of several brands of pollen substitutes and it was seen to be desirable to check these in the laboratory for their protein content.

The costs of the protein in the various forms of pollen substitute were compared. Kra-waite was by far the most expensive and Professor Hydak's formula based on locally available materials was the cheapest. It was composed of the following:-

Soya flour	4 parts by weight
Dried brewer's yeast	1 part by weight
Dried skim milk	1 part by weight

Honey, or 2 to 1 sugar syrup, is added to the mixture to produce a stiff paste. A half-pound cake of this can be spread over the top bars of the brood nest and covered with grease proof paper.

1968　　　DETERIORATION OF THE COLOUR OF HONEY

J.W. White and others had published a valuable paper on how processing and storage effect honey quality with particular emphasis on the damage done to diastase and invertase in honey and the production of 5-hydroxymethylfurfuraldehyde (HMF).

I needed to determine the effects on the colour of honey of heating to various temperatures during extracting and processing. Different types of honey were subjected to temperatures from 43°C to 80°C. The reading on the Pfund scale was checked regularly for up to one week and the results tabulated to show the relationship between time, temperature and colour deterioration.

The deterioration of the colour in storage is also a matter of great concern. Earlier work showing the rate of darkening per month in terms of increase in millimetres in the Pfund scale was tabulated for storage temperatures 10° to 37.8°C. The rate of deterioration was rapid when the temperature of the honey was above 30°C, particularly in galvanised drums stored in the sun.

This work was described in greater detail in Chapter 12.

References:

SMITH, F.G. (1967) Deterioration of the colour of honey *Journal of Apicultural Research* 6(2): 95-98

WHIYE, J.W. et al (1963) How processing and storage affect honey quality *Gleanings* 91(7): 422-425

1967　　　　　CAPPINGS REDUCER: A NEW DESIGN

I designed a cappings reducer, modified for use with circulating hot water, and the prototype was tested by a beekeeper in full commercial field use and worked most successfully. A few minor improvements were added to the specifications for future models.

Following the publication in the *American Bee Journal* of an article on the new cappings reducer, there was a steady demand for plans and specifications. Most requests came from America (34), followed by Canada (5), Australia (4), New Zealand (2), Mexico (1) and England

(1) – a total of 47 in about six weeks.

See also Chapter 13.

Reference:

SMITH, F.G. (1966) Cappings reducer: a new design *Amer. Bee J.* 106(9):333-335

1967-8   HOT WATER CIRCULATING SYSTEM

To overcome another defect in honey extracting plants, experiments were carried out to develop an effective straining device which would clear froth, particles of wax and other debris from honey as it flowed from the extractor.

This was achieved by means of a small baffle tank with replaceable strainers constantly immersed in the honey. It was applicable to both mobile extracting plants and central plants.

Details were described in Chapter 13.

Reference:

SMITH, F.G. (1968) A honey sump strainer *Apic.* 2:106-109

1969   HONEY PLANTS IN WESTERN AUSTRALIA

The above, Bulletin No. 3618, was published in 1969 by the Department of Agriculture.

This bulletin of 78 pages contains descriptions of the bee forage zones, a monthly calendar of bee plants, and illustrated descriptions of the main honey plants – 45 species of Eucalyptus. The identification of the species is based on the shape and arrangement of bus and fruits. The bulletin was designed to be of value not only for the beginner who might wish to keep bees in the traditional beekeeping areas of the State, but also for the experienced beekeeper who might wish to explore the potential of the very large and virtually untapped Mallee zone.

This book was planned five years earlier in consultation with the Executive Committee of the Beekeepers' Section, Farmers' Union of WA. Compiling material for it proved a formidable task and, as the work proceeded, it was found that there were many aspects on which fundamental research was needed.

Bulletin No. 6318 covers the southern part of the State which has an average annual rainfall of more than 230mm, and in which is concentrated all commercial honey production.

The descriptions of bee plants are confined to the Eucalypts and are illustrated by C.A. Gardner's excellent drawings and supplemented by others by Stan Chambers.

The following introduction to the bulletin explains some of the rationale behind its production, and that of the vegetation maps which followed in later years.

Successful honey production depends, among other things, on a good knowledge of the plants which produce nectar.

Every apiarist needs to know which plants are of importance to honeybees, where those plants occur, and when they flower. He also needs to know which plants produce nectar which will result in the production of good quality honey, and which produce unpalatable or unmarketable honey. To maintain the strength of his bee colonies he also needs to know which plants produce nutritious pollen.

The object of this bulletin is to provide the basic information on these subjects in the main beekeeping areas of Western Australia. The bulletin does not pretend to say everything that there is to be said on the matter as this would be quite impossible. Every year each beekeepers learns a little more about the honey flora, but no one year is exactly the same as the previous or any other year because flowering behaviour and nectar production are dependent on that infinitely variable factor, the weather.

A successful apiarist is one who has the aptitude for working with nature, and who is sensitive to the changing conditions which affect the flowering of plants and the behaviour of bees. In Western Australia, where the greatest honey production is produced by moving the bees from honey flow to honey flow, the apiarist has to keep himself informed on the condition of flora over very large areas of country. It is this constant contact with nature, and the study of the interplay of climate, soil, vegetation and bees, which is one of the great fascinations of beekeeping.

The response of the apiarist to nature's ever changing conditions is an art rather than a science. Mastery of this art should be the aim of everyone who owns a hive of bees. This bulletin outlines some of the science on which the art is founded, and possibly a little of the art itself.

Reference:

SMITH, F.G. (1969) *Honey plants in Western Australia* Perth: Dep. Agr. Bull. 3618

1972   VEGETATION MAP: PEMBERTON & IRWIN INLET
1973   VEGETATION MAP: BUSSELTON & AUGUSTA
1974   VEGETATION MAP: COLLIE

During the winter of 1970 I felt that I had run out of inspiration for further articles for *Apiculture.* I noticed that I had been repeating old themes so I decided that I should stop publishing immediately rather than let the standard slide, as I had see happen so often to good publications. The production of *Apiculture* during the past seven years was very much a one-man show, though Stan Chambers was a very co-operative helper. I was disappointed that the beekeepers themselves had not contributed articles from their own great wealth of experience; maybe they were just too busy trying to earn themselves a living.

When I discussed this with the Director of Agriculture, he agreed with me and suggested that I added mapping of the vegetation of the south west of WA to my responsibilities, with suitable recompense in upgrading my position.

I accepted the offer and the above three maps at the scale of 1:250

000 were the result of my efforts from the end of 1970 until I took up a new position as Director of National Parks in February 1974. Each map involved about one year's work in addition to my duties as Officer-in Charge, Apicultural Branch.

The vegetation maps were compiled and drawn in accordance with the requirements of the Western Australian Vegetation Survey Committee.

My primary sources of information were aerial photographs and the Forests Department Air Photo Interpretation (API) Plans which provided some additional information of vegetation structure and principal tree species occurring in forested areas.

I made traverses by motor vehicle, on foot and occasionally by boat, to check my interpretation of the vegetation structure and plant associations shown in the aerial photographs, which I had examined under a stereoscope in my office before starting each field trip. The routes covered are shown on the border of each map. My wife, Joan accompanied me on each of the field trips, which involved a full week each month. She did the driving while I compared the actual vegetation with my interpretation of the aerial photographs and recorded details of the plants communities.

I identified plant material in the field or, in the case of unfamiliar material, had it named at the Western Australian Herbarium.

I took and processed photographs to illustrate the text in a booklet which described the various plant communities and vegetation systems which occur on each map. Each photograph was taken at a place within the boundaries of the map and the location was given by map-grid reference.

I mapped the vegetation on the basis of structural criteria of the tallest stratum. Structural formations were illustrated by colours. Sub-divisions of these formations were on the basis of plant associations indicated by means of symbols.

The vegetation detail was prepared for printing by the Lands Department over the blue (rivers and lakes) and black (roads, place names and grid) base maps of the standard topographical 1:250 000 sheets in the Australian National Map Series. This made locations easy to find. The vegetation detail replaced the contour information of the standard maps.

Each of the maps with its booklet was published by the Western Australian Department of Agriculture.

CHAPTER 21

THE FIELD FOR IMPROVEMENT IN THE TECHNIQUES OF HONEY PRODUCTION AND EXTRACTION

Early in 1972, I was invited to address the First Australian Bee Congress which had been organised at Broadbeach on the Gold Coast of Queensland. In that talk, given on 13th October 1972, I expressed my concern at what I still saw as the major weaknesses in the Australian beekeeping industry, and I made my recommendations for overcoming those weaknesses.

The following is the text of that address:

*   *   *

Honey production in Australia is extraordinarily efficient. Had it not been, the industry would not have survived the long period of low prices which we have suffered. Nevertheless there is room for improvement in both the quantity of honey produced per hive as well as in the quality of the honey reaching the consumer.

The aspects on which I propose to speak will not only achieve greater efficiency, but will also make the work of honey production more pleasant for the beekeeper and for his assistants.

There is room for **improvement of stock.** The selection of breeding stock should be based on performance, not on appearance. Colour is of importance only in indicating possible mis-mating. Nevertheless we still find some beekeepers wanting golden Italians, in spite of generations of experience that pretty bees are often useless bees. Queens cannot be judged by appearance alone, they are not beef cattle or porkers. Queens can only be judged by the performance of their progeny

In the selection of breeding stock, the aims are high production and good management characteristics. The bees must be hard working. The queens must be good egg layers, yet the colony must show economy with food consumption. The colony needs to build up before the honey flow, not on it. The bees should show resistance to disease. There should be a disinclination to swarm; bees prone to swarming are useless to modern beekeeping. The workers need to have a long life; there needs to be a balance between fecundity and long life. The bees need to be good defenders of their hives; this must be balanced by manageability. There must be economy in the use of stores during periods of dearth. There must be readiness to build comb. The arrangement of stores in the hive is important; pollen should be seen to be stored around the brood nest and honey in the supers. Bees that spread their brood nests up through three of four boxes make management difficult. Good temper and calm behaviour are vital to management. Disinclination to use propolis or brace comb also assists management. Even and level cappings on honey comb aid uncapping.

Once a beekeeper has a good strain he should do his own breeding, selecting bees most suited to the environment in which he works and suited to his form of management.

Much honey is lost through **overstocking apiary sites.** Too many apiaries in an area are an obvious cause of loss; beekeepers usually avoid this. Too many hives in a single apiary is a frequent cause of a poor honey crop. During the active season each bee colony uses about 700g of honey each day for its own maintenance. Each colony has to gather this amount to keep it going. Any additional honey or nectar gathered is surplus: this becomes the beekeeper's crop.

Assuming that all hives are in condition for foraging, the amount gathered each day will be dependent on how much nectar is available within economic flight range. If there is the equivalent of 70kg of honey available to be gathered within range, 100 hives of bees will obtain only just enough to keep them going; there will be nothing for the beekeeper. Fifty hives of bees in such a place will gather 35kg for themselves and 35kg for the beekeeper. Twenty strong hives will gather 14kg for themselves and 56kg for the beekeeper each day.

To get the best results, each beekeeper needs to be able to assess the probable daily yield of nectar from the area around each apiary site. This is learnt only from experience. The beekeeper should then stock each apiary with sufficient hives to give him the maximum yield of honey in the form of crop. This should be on the basis of a surplus of between 136g and 272g of honey per day per hive.

Beekeepers accustomed to handling hives of bees by the truck load find this concept difficult. But stocking each apiary with only as many hives as will give the maximum return to the beekeeper is essential for the most profitable honey production.

Combs used for the storage and extraction of honey should be used for that purpose alone. The frames to hold the combs should be designed for maximum efficiency in uncapping and extracting. Supers in which honey is stored should be of a size which is easy to handle when full of honey, not too heavy, not too light.

The Dadant depth shallow super, 168mm deep and in the Langstroth 10-frame width, meets these requirements. This is referred to locally as the Manley super. When full it averages 25kg gross, ranging from 23kg to 28kg. The amount of honey in it averages 18kg, ranging from 16kg to 21kg; three-quarters of that held by a full-depth Langstroth ten-frame box. Many beekeepers have recognised that the full depth 10-frame box is too heavy as a honey super, averaging 36kg gross; it may be up to 41kg. The more manageable weight of the Manley super increases the speed of handling, it lessens fatigue, and eliminates strain and backache from handling full-depth boxes. The bees fill and ripen the honey in theses Manley supers more quickly. They are more quickly and thoroughly cleared of bees by Porter escape clearer boards, or by phenol or other repellents or by air blowers.

Manley frames are of simple construction. They are designed for the storage and easy extraction of honey, they are self-spaced at the best spacing for honey supers, they do not crush bees when moving hives with empty supers, there is no fiddling about spacing the frames by hand, and boxes can be stood on ends or sides without crushing bees or comb. The width of the top and bottom bar is just right for the removal of cappings with one clean sweep of the knife. Very little honey is cut off with the cappings. Less honey has to be separated from the wax; less honey is damaged in cappings reducers.

The essential features in the design of Manley frames are that the end bars, which are parallel sided, are 16mm wider than the top and bottom bars. The top and bottom bars are both the same width as each other for ease of uncapping. This 16mm difference between the width of the end bars and the top and bottom bars is vital. The bees cap honey comb when there is 12mm to 13mm between the faces of adjacent combs. The design of the Manley frame permit the cappings on the combs to stand proud of the top and bottom bars by 1.5mm to 2mm so that they can be uncapped easily with one stroke of the knife.

For use in Australian 10-frame size honey supers, the width of the top and bottom bars of Manley frames should be 27mm and the width of the end bars 43mm. Eight of these frames go into a 10-frame super, allowing ample space at each side for expansion of the wood in the humid atmosphere of the hive and for easy removal of the frames. The depth of these frames is 159mm.

For the hand operator, and that includes most beekeepers, Manley frames are uncapped in half the time taken to uncap full-depth frames, with a shorter knife and much less fatigue. The use of 168mm Dadant depth supers and Manley frames results in faster handling and uncapping, a decrease in fatigue and strain, and more honey can be extracted each day than with full-depth supers. As the combs are not used for brood production, the honey is free from the taint of dark combs, very little slumgum is produced in cappings melters and better quality honey results.

Because of the design of Manley frames, very little honey passes through the cappings reducer. Fuel is saved because less heat is taken up by honey and the maximum amount of honey is extracted in the proper place, the honey extractor. Finally, reduction in slumgum permits better flow of heat to the wax and results in better quality beeswax.

Central **honey extracting plants** are seen to be preferable to the use of mobile field extracting plants. Field extraction is laborious and time-consuming. The bees in the apiary are subjected to prolonged disturbance and the extraction has to be carried out while the honeyflow is in progress; at any other time serious robbing occurs. This frequently results in the extraction of freshly gathered nectar. Field extraction requires long absences from home and help is not easy to obtain.

Central extraction enables the design of more efficient extracting plant. The apiary work becomes more efficient; there is least disturbance in the apiary and reduced labour in the apiary. The robbing risk is reduced and the crop can be harvested after the end of the flow. There is better home life and more pleasant working conditions. Staff becomes easier to obtain and casual labour can be obtained as required.

Circulating **hot water systems** have a number of advantages over the use of steam. Steam has long been used to supply heat for the honey extractor, the uncapping knife and the cappings melter, also for cleaning down equipment. Steam has the disadvantage that it is of a higher temperature than that needed to do the work. Heat is therefore wasted, fuel is wasted, working conditions in the plant are hotter than desirable and, unless a condensing system is incorporated, water is wasted.

Hot water circulating systems have the advantages of providing heat at precisely the temperature required to do the job. Residual heat is not wasted as the water recirculates through the system.

Working conditions are more pleasant. Damage by local overheating of honey and beeswax is eliminated. The overall reduction in heat losses saves fuel, and water losses are minimized.

Draining honey from cappings has advantages over the traditional steam-heated cappings melter. The steam-heated melter has long been seen to be the source of damaged honey; up to 25% of the crop is so treated in some plants. The use of Manley honey-storage frames reduces this proportion considerably. But the aim should be to eliminate all damage to honey.

The causes of damage to honey are local overheating when steam is used, chemical reaction with metals at high temperature, leakage of steam and water into honey and absorption by honey of the soluble stains in darkened cappings and pieces of comb. Even if the first three causes of damage are eliminated by the use of hot water, stainless steel and sound plumbing, the melting of beeswax in contact with honey results in the transfer of dark pigments from the wax to the honey.

Honey should be allowed to drain from the cappings in baskets. This is usually considered as a bulky and slow operating arrangement, but there are compact refinements to this system with a series of baskets which drain in a warm cabinet.

After draining during the day, the wax can be melted by turning up the heat overnight, or by transferring the cappings to a separate beeswax rendering tank. Draining by gravity from a series of baskets presents fewer problems than do centrifugal extractors, although some have been used successfully.

The usual type of **honey sump,** receiving honey flowing from the extractor, frame rack and cappings drainer, mixes air and wax particles into the honey by allowing the honey to fall into the sump. To avoid beating in air and to aid the separation of air and wax from the honey, the honey should enter the sump at a level well below the surface, and be flowing horizontally. The light wax and air can then rise to the top and the clear honey settles to the bottom.

A baffle should be provided to maintain the honey level constant in the inlet chamber and to draw off the honey only at the bottom, thereby holding back the froth and wax. A strainer should be immersed in the honey in the sump. A strainer held above the surface quickly clogs and beats air bubbles into the honey. A simple method of straining is to insert a close-fitting flat screen between a pair of baffles. Screens can be changed easily by putting another screen behind the first and withdrawing the clogged screen. For larger operations, immersed hanging basket strainers can be used in the middle section of the sump. For practical purposes we have found that 32 mesh screens with baffles provide adequate honey clarity. A screen of 32 mesh is one having 32 wires per inch in each direction. The honey must go through the wax and air separation section before being allowed to flow into the basket. It should not be poured direct from the extractor into the basket. The third section of the sump behind another pair of baffles contains the pump float switch and the outlet. The honey is drawn from the bottom and the baffle maintains the level in the straining section. Arrangements are required to enable each section to be drained at the completion of an extraction.

The introduction of **honey drums** having one end which can be completely removed has greatly improved the decanting of honey in the packing plant. If the honey has granulated in closed–ended drums, prolonged heating is required to liquefy honey in the 200-litre drums before it will flow out the bung hole. This is the major cause of damage to the colour of honey in packing plants. Open-ended drums enable a very short heating time to be used to free the solid plug of granulated honey. Once clear of the drum, this plug can be broken up by passing over heated tubes and by mechanical means, before being pumped through a heat exchanger to complete the dissolving of the crystals.

In my opinion, the **production of honey can be increased** by improvements in the selection of breeding stock and by stocking apiaries with the number of hives suited to the availability of nectar in the area.

Improvement in the quality of honey produced, as well as in working condition, can be obtained by –

1. The use of Dadant depth supers with Manley frames,
2. The establishment of central extracting plant,
3. The use of circulating hot water systems instead of steam,
4. Using draining baskets for separating the honey from the beeswax cappings,
5. The use of a simple but efficient honey sump, and
6. The use of open-ended drums instead of closed-ended drums.

*   *   *

This was my last major activity in the field of beekeeping extension work apart from editing the scientific papers presented at the First Australian Bee Congress, before they were published by Apimondia. I was by this time very involved in the production of vegetation maps of the south west of Western Australia and this work ceased only when I took up the duties of Director of National Parks.

At first sight, the transfer from apiculture and vegetation mapping to National Parks seems a pretty long jump, but in fact it was not: the administration of a Branch and the practical work of mapping had been valuable training in addition to my previous experience in the British Army and the Colonial Service. I knew many of the parks which, at that time, covered a total area of nearly 2,000,000 hectares, and had studied their vegetation. By the time I had retired it was about 4,000,000 hectares.

In National Parks my Aberdeen training for a forestry degree really came into its own, and the next years were a time of tremendous interest and creativity, defining a management policy, preparing management plans for individual parks, planning roads 'to lie lightly on the land', getting rid of some of the worst mistakes that had been made through ignorance, and increasing and welding together a splendid body of rangers into a united and enthusiastic brotherhood, highly regarded by visitors to the parks. It was a great sadness to retire and leave them.

FRANCIS G. SMITH

APPENDIX

A METHOD OF RESEARCH INTO BEE FORAGE

Early in 1983 I went to Thailand at the request of the Australian Government to advise the Thai Department of Agriculture on apicultural research. One outcome of that visit was a request from my Thai colleagues for details of the method I had used in my bee botany research in Africa and Western Australia. As the only record of that information was in the thesis for my Doctorate, I am reproducing it here with minor amendments.

Introduction

Before one can begin to consider methods of bee management, a sound knowledge is required of bee behaviour and bee forage. Indeed, these two subjects are the foundations on which apiculture is based.

The behaviour of bees and their reactions under various circumstances has been, and still is, under investigation by a number of very capable workers; there is little point at the present time in duplicating their investigations. However, no systematic work on bee forage has been done which is relevant to South East Asia or to many other parts of the world. It is, in fact, a subject on which little is known other than the few scraps of information gleaned by some of the more observant beekeepers.

Before starting serious work on the actual management of bee colonies, the major task to be accomplished by professional apiculturists is a thorough study of bee forage, in particular in areas which appear to have the greatest potential for beekeeping.

The study of bee forage, sometimes call bee botany, is the study of those plants which are of importance to honey bees. Bees are dependant for their very existence on flowering plants which provide them will all their food in the form of nectar and pollen. It is essential for the beekeeper, who wants to manage his bees for the production of maximum crops of honey and beeswax, to know which plants are of value to bees, where they occur, and when they flower. He needs to know which produce only subsistence for the bees and which produce heavy yields of nectar thereby enabling the bees to store surplus food. Also, he needs to know which plants result in the production of high quality honey and which cause unmarketable honey to be produced.

The best way of getting most of the information required is from the bees themselves. When bees collect nectar they also obtain a small proportion of pollen from the flower. This pollen has either fallen, blown or been knocked into the nectar or clings to the hairs of the bees. After the nectar has been converted into honey in the hive, the pollen remains present in the honey, which is an excellent preservative. In addition, bees collect pollen for food and store it in the combs.

If the pollen in the honey, and that which is stored in the combs can be identified as to the plants of origin, we will then know which plants are of value to the bees. The relative importance will depend upon the quantities of honey collected containing particular pollen grains, not on the actual quantities of pollen, because the honey from different plants contains pollen in different proportions.

To identify the plants from which pollen has come, it is necessary to build up a collection of named pollen preparations on microscope slides. The pollen grains forming the collection of standards must be collected from correctly identified plants. Thus a major part of this work is the task of collecting pollen samples and herbarium specimens from as many plants as possible, having the herbarium specimens named, preparing and mounting the pollen samples, and describing the pollen grains of each species, recording that description in such a manner that it can be picked out from the whole collection easily.

The next stage is the collection of honey samples and hive material from different areas and at different seasons, and the identification of the plants of origin as indicated by the pollen grains extracted from the samples.

The field notes made at the time of the collections of the herbarium and pollen specimens, knowledge of the composition of different plant communities, together with the identification of the plants useful to bees, provide the information which is so essential for enlightened management of bee colonies.

POLLEN MORPHOLOGY

The walls of pollen grains reveal a variety of structural features. There may be apertures in the walls. These may be circular or elongated. The walls are composed of different layers and the apertures in the inner layers may be of a different shape from those in the outer layers. The walls often show markings due to the uneven nature of their surfaces. The layers of the walls may be of different thicknesses. Pollen grains have a variety of different shapes and sizes.

The normal optical microscope, with good quality lenses, enables one to examine the features in the walls of the pollen grains. As the characteristics of pollen grains of any species of plant are more or less constant, it is possible to use these features in identifying the plants from which the pollen has come. In a few cases the pollen grains of a single family, or even of different families are so similar that one cannot do more than name the family or groups of families. Usually however, genera or groups of genera are distinguishable and very often species can be named with certainty.

Pollen grains may be symmetrical in shape or may be irregular. Normally they are symmetrical. The grains are regarded as having a **polar axis.** This coincides with a line through the centre of the grain to the centre of the tetrad of grains in the mother cell. The proximal pole is the centre of the surface nearest the centre of the tetrad; the distal pole is the centre of the surface furthest from the centre of the tetrad. The majority of the pollen grains are found in the free state, separated from the tetrad, but it is useful to bear in mind this concept of a polar axis when describing the nature and position of features.

Some pollen grains have no distinguishing features to enable the position of the poles to be determined. These are termed **apolar. Isopolar** grains are those in which the proximal and distal poles are similar in appearance. In **heteropolar** grains the markings at the poles are different.

The **equatorial plane** cuts the pollen grain at right angles to the polar axis. Vertical planes pass through the polar axis. If more than two vertical planes are symmetrical, or if only two are symmetrical but their equatorial axes are equal, the grain is termed **radiosymmetrical.** If there are two vertical planes of symmetry with unequal equatorial axes, the grain is **bilateral**.

In the radiosymmetrical grain the length of the polar axis and the equatorial diameter may be equal. Such a grain is **spherical** or **spheroidal**. On the other hand, one or other may be longer. A series of terms is used to describe the shape in equatorial view according to the proportional lengths of the polar axis and equatorial diameter. They are as follows:-

Polar Axis : Equatorial Diam.	Shape Class
Less than 4:8	Peroblate
Between 4:8 and 6:8	Oblate
Between 6:8 and 7:8	Suboblate
Between 7:8 and 8:8	Oblate spheroidal
Between 8:8 and 8:7	Prolate spheroidal

Between 8:7 and 8:6	Subprolate
Between 8:6 and 8:4	Prolate
Greater than 8:4	Perprolate
8:8	Spherical
Between 7:8 and 8:7	Spheroidal
Between 6:8 and 8:6	Subspheroidal

In the descriptions of pollen grains the measurements quoted (in microns) are the diameter in the case of spherical grains, the length of the polar axis (ap) and equatorial diameter (ed) of other radiosymmetrical grains, and the polar axis and the maximum and minimum dimensions in the equatorial plane of bilateral grains. The size of outstanding features such as length of spines, height of ridges and width of apertures may also be stated. In the case of grains having spines on the surface, the measurements are made between the main surfaces, the length of the spines not being included in the polar axis or equatorial diameter.

APERTURES

Most pollen grains have apertures or openings in the hard wall of the grain. The opening may be covered with a thin membrane which may be smooth (**psilate**), **granulate** or **crustate** (having coarse granules close together).

If the apertures in the outer layers of the wall are circular they are termed **pores**. If they are elongate they are termed **furrows**. If the apertures in the inner layers of the walls do not coincide in size or shape with those in the outer wall, the apertures are termed **orate**. These inner apertures (**ora**) may be circular or elongate. If elongate, they may be longer in the horizontal plane, i.e. latitudinally elongate, and are called **lalongate**; or they may be longer in the vertical place, i.e. longitudinally elongate, and are called **lolongate**. If they are continuous, running into each other in the horizontal plane, they are termed **zonorate**.

Apertures arranged round the equator of the pollen grain are said to be **equatorial**. If they are scattered evenly over the surface of the grain they are **global**. Apertures occurring at the proximal pole are **proxipolar**, at the distal pole, **distipolar**; at both poles, **bipolar**.

Tetrad scars occur on the proximal pole. If elongate, these are termed **laesurae**. A pollen grain with one laesura is termed **monolete**. If the tetrad scar consists of three arms radiating from the pole it is **trilete**; if it is a more or less circular aperture, the grain is **hilate**.

A furrow passing through the distal pole is a **sulcus**. The pollen grain is **sulcate**. A circular aperture at the distal pole is an **ulcus**.

Global pores are sometimes called **foramina**. Global furrows are **rugae**. Grains having foramina are **forate** and those with rugae are termed **rugate**.

Furrows crossing the equator at right angles are **colpi**. Pores arranged round the equator are **pori**. Thus we have **colpate** and **porate** pollen grains.

The area bounded by two adjacent colpi and two lines joining their ends is the **mesocolpium**. The areas outside the mesocolpia are the **apocolopia**. These may be polar or equatorial.

In polar view, optical equatorial section, it may be seen that the apertures, colpi or pori, are situated in ditch like depressions; such a grain is **fossaperturate**. Or they may be located on flattened surfaces, in which case the term **planaperturate** is used. Should the apertures be on protruding angles of the walls, the grain is **angulaperturate**.

In the case of fossaperturate and planaperturate grains, the mesocolpia are convex. Angulaperturate grains may have mesocolpia which are convex, flat or concave.

Colpi may meet at the poles, in which case the grains are **syncolpate**. If the colpi branch at the ends and the branches of adjacent colpi meet, the grain is **parasyncolpate**. If, on both sides of the equator, there are sets of colpi which, if they met at the equator, would form ordinary colpi, the condition is described as **diplodemicolpate**. If the demicolpi of the proximal face meet at the proximal pole and those of the distal face meet at the distal pole, the pollen grain is said to be **syndemicolpate**. If the colpi are arranged in pairs, the grain is **geminicolpate**. If there is one os in each aperture, it is orate. But if there are two it is **diorate**. Cases occur when some furrows are orate and others have no ora. If the furrows are alternatively orate and non-orate, the term **demiorate** is used. A grain having furrows which are more or less colpi-like is **plicate**.

An **aspidate** pore is one surrounded by a circular ring of thickened tissue. Should the apertures be arranged in the form of a spiral, the grain is termed **spiraperturate**.

ORNAMENTATION

Under the optical microscope, the wall or shell of the pollen grain may appear to be composed of one or several layers. The hard outer layers, which contain the ornamentation, are called the **exine**. The inner soft unornamented layer is known as the **intine**. Erdtman (1952 b) subdivides the exine into the outer sexine or sculptured part of the exine and the inner **nexine** or unsculptured part.

When distinguishable, the exine seems to be composed of rods (**pila**) in the sexine layer arranged at right angles to the surface of the nexine. These pila have swollen heads supported on pillars. When the heads touch, the surface is smooth (psilate) but when there are minute vertical holes it is termed **punctate**. On the surface there may be granules (**granulate**), warts (**verrucose**) or spines (**spinose**). If the heads of the pila are free, the ornamentation is termed **pilate**. Often the heads are joined together forming patterns. If they take the form of islands separated by grooves the term used is **areolate**. If they are in the form of a network of walls surrounding spaces, the pattern is said to be **reticulate**. **Lophate** grains have very high walls and these may be **psilolophate** if smooth on top or **spinolophate** if they have pines on top of the walls. If the pattern is in the form of parallel lines it is **striate**; if crinkled it is **corrugate**. Grains provided with airsacs are **saccate**.

This terminology has been limited to that which was found useful in earlier studies of bee forage (Smith 1956). For the most part it follows the recommendations of Erdtman (1952 a) with minor modifications. Erdtman (1952 b) gives a more extensive terminology.

COLLECTION OF MATERIAL

The pollen grains may be collected from growing plants or from herbarium material. Herbarium material may be contaminated with pollen grains from other plants. Collecting from growing plants makes possible the collection of important data on the actual conditions under which the plants grow, when they flower and, if collected in the wild, the plant community with which they are associated.

Pollen bearing material from flowers is put into vials up to 50mm long by up to 9mm in diameter containing about 1ml of glacial acetic acid. The acid resistant caps are numbered.

Sufficient plant material is collected for sending to the official herbarium for naming together with some spare for a reference sheet to be retained in the laboratory.

I found that it was most convenient to enter the field notes on a record card (200mm by 125mm) which I subsequently filed in numerical order under the collector's name. The collector enters on the front of the card the date, his or her name, specimen number, local name of the plant if known, and a description of the habitat and locality. On the reverse of the card are entered such features of the botanical description as are required to add to those apparent in the herbarium specimen. When the specimen has been named, the genus, species, author and family are added to the front of the card.

The vials are most conveniently carried in a small wooden box having cardboard dividers. The herbarium specimens should be placed between sheets of drying paper in an herbarium press as soon as collected, together with the specimen number. The drying sheets should be changed daily and the damp sheets spread out to dry. It is more important to transfer the specimen number with the herbarium material. In moist areas the drying papers may need to be changed more frequently to prevent the development of mildew.

The record cards I used I had designed for sorting but in practice I found that this was unnecessary. But it is convenient to have them printed in a standard form so that no information is overlooked when recording the collection of pollen grains and herbarium specimens in the field.

Back at the research station, as soon as the material is dry, an herbarium sheet is prepared and numbered and the remainder of the herbarium material is sent with appropriate field notes to the official herbarium for naming. It may be found that the names come back from the herbarium more quickly if small batches of herbarium specimen (not more than ten) are sent at any one time. Large collections tend to be set aside until a botanist has time to spare.

The pollen grains are treated, a microscope slide of the grains prepared and the grains described. The pollen description is entered on a specially designed pollen description card having the Paramount sorting system. The original collector's name and reference number appears on the field record card, the herbarium sheet, the pollen slide, the pollen description card and, if a photograph is taken, on the negative index.

PREPARATION OF POLLEN MATERIAL

I tried various methods of preparing pollen slides, in particular those recommended by Wodehouse (1935), Maurizio (1951 and 1953), Deans (1951) and Erdtman (1943 and 1952 b). At the end of all my trials, I found that Erdtman's method, although more complicated, produced the most satisfactory preparations for the examination of structural detail. Two outstanding features of this method were the removal of the plasmic material which otherwise distorted the light passing through the grain, and the staining of the wall of the grain a yellowish brown which made it easy to examine in pale blue light. Previously I had tried the whole range of microscope stains recommended by Wodehouse and others but none gave such satisfactory results as Erdtman's acetolysis method.

The method used is as follows:-

1. Number sufficient 15ml graduated conical centrifuge tubes for the day's work.

2. Shake each vial to get the pollen material in suspension in the glacial acetic acid, and using the cap to hold back the anthers and other large particles, decant acetic acid and pollen into a numbered centrifuge tube.

3. Record the number of the pollen vial and that of the centrifuge tube.

4. Add glacial acetic acid as required to balance the centrifuge tubes in pairs.

5. Put each pair of tubes into opposite buckets, centrifuge, and decant the acetic acid to leave the pollen grains in the tubes.

6. Prepare the acetolysis mixture by adding one part of concentrated sulphuric acid slowly, drop by drop, using a burette, to nine parts of

acetic anhydride. Prepare only sufficient of this acetolysis mixture for one day's work. Do this under a fume hood or in a very well ventilated laboratory.

7. With a graduated 5ml pipette with a rubber bulb, add 5ml of this acetolysis mixture to each centrifuge tube.

8. Place a clean glass stirring rod into each tube and transfer the tubes to a water-bath at room temperature.

9. Heat the water to boiling point, turn off the heat and stir the fluid in each tube.

10. Remove the stirring rods and centrifuge for about one minute.

11. Decant the acetolysis mixture into a beaker for subsequent disposal.

12. Add 100ml distilled water to the pollen in each tube and shake each tube well, taking care not to contaminate the contents. The foam that is produced is dispersed with two or three drops of acetone or alcohol.

13. Centrifuge and decant the liquid.

14. Add about 3ml of 1:1 mixture of distilled water and glycerine and leave for at least ten minutes. It can be left until the next day at this stage.

15. Centrifuge for at least two minutes and decant the liquid.

16. Place the tubes upside down on filter paper to remove the surplus liquid.

17. Add one drop of hot glycerine jelly to the pollen grains in the bottom of the tube, shake to distribute the grains throughout the glycerine jelly.

18. With a platinum loop transfer a very small amount of pollen-bearing glycerine jelly to the centre of a microscope slide, previously marked with the number of the pollen sample.

19. With one edge of a thin circular cover glass resting on the slide, and a needle under the edge on the other side, lower the cover glass on to the pollen-bearing jelly, and heat the slide gently over a spirit lamp to cause the jelly to spread evenly; the jelly should cover a more or less circular area with a diameter of about 5mm less that the diameter of the cover glass – it is essential that the jelly does not reach the edge of the cover glass. Leave the slide until the next day.

20. Seal the preparation with Canada balsam using a turntable or by applying a drop of melted paraffin was on a glass rod to the edge of the cover glass and heating the slide gently so that the paraffin spreads quickly in under the cover glass.

Note: Glycerine jelly (Kisser's method) –
    gelatine 50g,
    distilled water 175ml,
    glycerine 150ml,
    phenol crystals about 7g.

Pollen taken from the legs of bees and pollen taken from the comb is treated in the same way as pollen material collected from flowers, being placed first in vials of glacial acetic acid.

To make pollen preparations from honey, 20ml of honey are dissolved in 60ml of distilled water, warming if necessary. Pour 10ml of the dissolved honey into a centrifuge tube. Centrifuge for one minute and decant the liquid. Add another 10m to the same tube, centrifuge and decant. Repeat until all the dissolved honey has been centrifuged or it is apparent that there is sufficient accumulation of pollen at the bottom of the centrifuge tube. The amount of pollen present in honey varies very much. This variation is partly due to the structure of the flowers from which the nectar was collected and partly due to the method of extraction of the honey from the comb. To obtain the most accurate information as to the source of honey, the sample should be scooped directly from new (white) honeycomb which contains no stored pollen.

Add 5ml of glacial acetic acid to the pollen sediment and shake the tube well. Preparation then proceeds as previously described (steps 1-20 above).

DESCRIPTION OF POLLEN GRAINS

To examine the pollen grains a microscope is used with an incline binocular tube, a triple nosepiece, a rectangular mechanical stage, a focussing and centring substage, an achromatic condenser and built-in illumination. It is a great convenience if the binocular tube has a photo tube to which can be fitted a 35mm camera body. Paired compensating eyepieces with 10x magnification are most useful. Preliminary examination of the slides is done with a 10x objective but detailed inspection of the pollen grains is done with a 40x apochromatic objective. The illumination should have a pale blue filter. A 100x oil immersion objective may be used occasionally, but it is not essential for this work.

Measurements are done with a drop-in eyepiece micrometer disc. The value of the eyepiece graticules in thousandths of a millimetre (microns) with the 40x objective is determined by examination of a stage micrometer slide having 1mm with 0.01mm divisions. All measurements of pollen grains are given in microns.

The characteristics of the pollen grains are entered on the pollen description cards. After completion of the description, the holes opposite the relevant features are punched away. These cards were designed by the author to be sorted by the Paramount system. Provision is made for the cards to be picked out by the number of apertures, type and position of apertures, characteristics of furrows and ora, shape, ornamentation, radiosymmetrical shape classes, length or equatorial diameter and polar axis as well as by reference number or family. Originally when the cards were printed it was intended that they should be sorted by genera, but it was found to be more suitable to sort by families. Thus where the cards are marked Genus, the Family is punched.

As the collection of record cards increased, it was found necessary to introduce a preliminary sorting series. This was done by the preparation of family cards. Those families whose pollen grains fall into several distinct types had a card prepared for each type. On the family card are entered all the characteristics of all pollen grains collected in that family. Thus, when it is desired to identify a pollen grain found in honey or hive material, the family cards are sorted with the needle first. Those family cards which have the characteristics required fall out. Then each family whose family card has fallen out is sorted in turn. The result may be several pollen description cards of different families each having the characteristics sought. The next stage is to examine under the microscope the slides corresponding to the cards until the pollen slide corresponding in all detail to the specimen to be identified is found. This part of the work is simplified by the use of photographs of pollen grains.

Mention should be made of the variations which occur. Pollen size appears to be liable to variation between localities and individual plants. There may be considerable variation in the size of the grains produced by a single plant. Variation may also occur in the shape classes in the radiosymmetrical series. This variation can be due to mechanical deformation under the pressure of the cover glass on the microscope slide. This applies also to the shape and width of furrows. A pollen grain, the polar axis of which has been compressed, will display very much wider furrows than will appear in equatorial view. The measurements which are recorded should be those of grains which are believed to be undeformed.

ILLUSTRATIONS

Photographs are used to illustrate important features of the pollen grains described. As the depth of focus is very shallow, a photograph shows only an optical section through the grain. While this has the disadvantage of not showing the grain as a whole, it reveals structural details difficult to describe in words.

The photographs are taken with a 35mm camera body mounted on a phototube above a projective eyepiece. It is useful to use the first frame on each film to photograph the state graticules marked in 0.01mm so that the negative obtained can be enlarged to a constant final magnification of 1000x natural size.

A fine grain, high contrast film is used, such as Ilford Pan-F or Kodak Panatomic X. The initial film should be used for making test exposures of slides of pollen grains having different densities and sizes. A range of exposures from 1/25 second to 1 second should be tried. Pollen grains of different species range in size from about 10 microns to at least 140 microns in diameter, so at least two projective eyepieces are required. Those which enlarge the 40x objective image 4x and 8x have been found most useful. The 8x compensating projective was the most used and the 4x was used for the largest pollen grains.

References:

Deans, A.S.C. (1951) *The pollen analysis of honey* Lecture: The Central Association of Beekeepers, Ilford.

Erdtman, G. (1943) *An introduction to pollen analysis.* Verdoorn, New Ser. Pl. Sci. Books 12 Waltham, Mass.

Erdtman, G. (1952 a) On pollen and spore terminology. *Palaeobotanist*, Lucknow 1: 169-176

Erdtman, G (1952 b) *Pollen morphology and plant taxonomy: Angiosperms (An introduction to palynology 1)* Almqvist & Wiksell: Stockholm

Maurizio, A. (1951) Pollen analysis of honey. *Bee World* 32 (1): 1-5

Maurizio, A. (1953) Report of the I.U.B.S. International Commission for bee botany 1952 *Bee World* 34 (3): 48-51

Smith, F.G. (1956) *Bee Botany in Tanganyika* D.Sc. Thesis: University of Aberdeen

Wodehouse, R.P. (1935) *Pollen grains; their structure, identification and significance in science and medicine* McGraw-Hill: New York and London

PUBLICATIONS
by F.G. SMITH

1. SCOTLAND

(1948) Hive and frame problems *Scot. Beekeeper* (6):110

(1949) Timber for hives *Scot. Beekeeper* (2):26

2. TANGANYIKA

(1951) *Preparation of honey for sale* Tanganyika Agric. Lflt. No. 17

(1951) *Sababu ya kutowaangamiza Nyuki* Tang. Agric. Lflt. No.18

(1951) *Modem methods of rendering beeswax* Tang. Agric. Pamphl. No. 49

(1951) *Notes for beekeepers in Tanganyika* Tang. Agric. Pamphl. No. 50

(1951) Preliminary report on Trigona wax *E. Afr. Agric. J.* 16(4):185-187

(1951) Beekeeping observations in Tanganyika 1950/51 *E. Afr. Agric. J.* 17(2):84-87

(1952) Beekeeping observations in Tanganyika 1951/52 *E. Afr. Agric. J.* 18(2):59-61

(1953) Beekeeping in the Tropics *Bee World* 34(12): 233-245

(1954) Notes on the biology and waxes of four African Trigona bees *Proc. Ser. "A" Roy. Ent. Soc.* 29(4-6):62-70

(1954) *Kutengeneza nta ya biashara* Tanganyika Bee. Div. Lflt. No. 1

(1954) *Tumia bomba la moshi* Tang. Bee. Div. Lflt. No. 2

(1955) *Kutengeneza asali nzuri ya biashara* Tang, Bee. Div. Lflt. No. 3

(1955) *Beeswax* Tanganyika Bee. Div. Pamphl. No. 1

(1956) *A course of twelve lectures on beekeeping* Tang. Bee. Div. Pamphl. No. 2

(1956) *Honey* Tang. Bee. Div. Pamphl. No. 3

(1956) *Bee Botany in Tanganyika* D.Sc. Thesis, University of Aberdeen

(1957) Bee Botany in East Africa *E. Afr. Agric. J.* 23(2): 119-126

(1957) *Notes on buying beeswax* Beekeeping Div. Lflt. No. 4

(1958) The origins and functions of the Beekeeping Division *Empire Forestry Review* 37(3): 159-164

(1958) Beekeeping observations in Tanganyika 1949-1957 *Bee World* 39(2): 29-36

(1958) Foul brood in Tropical Africa *Bee World* 39(9): 230-232

(1958) Communication and foraging range of African bees compared with that of European and Asian bees *Bee World* 39(10): 249-252

(1958) Honeybees of Africa *African Beekeeping* (9-10)

(1958) The honeybees of the tropics *Indian Bee J.* 20(8): 108-112

(1959) The place of origin of Apis mellifera *African Beekeeping* (11)

(1959) Hives and management *African Beekeeping* (12-13)

(1959) The use of starters in brood frames *African Beekeeping* (14)

(1959) Beekeeping research in tropical Africa *E. Afr. Farmer and Planter* 3(10): 22-25, 35

(1959) Becoming bee-minded *Corona* 11(8): 306-308

(1959) *Beekeeping in Northern Rhodesia: its prospects and recommendations for its development* Tabora: Beekeeping Division pp. 21

(1959) *The Miombo in relation to beekeeping* C.C.T.A. Conference Paper

(1960) *Bee-keeping in the tropics of Africa* Lecture: Cent. Assoc. of Beekeepers

(1960) Breeding better bees *Rhodesian Beekeeping* 1 (1): 2

(1960) *Beekeeping in the Tropics* London: Longmans pp. 280

(1960) Comb foundation: its use for African honeybees *Bee World* 41(9): 235-240

(1960) The African Committee of the Bee Research Association *Rhodesian Beekeeping* 1(2): 1-2

(1960) Honey and beeswax production *Ukulima wa Kisasa* (63) 4pp.

(1961) *The African Dadant hive* Tang. Bee. Div. Pamphl. No. 4

(1961) Races of honeybees in Africa *Bee World* 42(10): 255-260

(1961) *A beekeeping research institute for Africa* Madrid: XVIII International Beekeeping Congress

(1961) *Improved brood frame for the African Dadant hive* Tang. Bee. Div. Lflt. No. 5

(1962) *Beekeeping as a forest industry* Dar es Salaam: Forest Dept.

(1963) *Beekeeping* London: Oxford Univ. Press pp.128

(1964) Some pollen grains in the Caesalpiniaceae of East Africa *Pollen et Spores* 6(l): 85-98

3. AUSTRALIA

(1963) *An Introduction to Beekeeping in Western Australia* Bull. Dep. Agr. W. Aust. 3108

(1963) *Foul brood diseases in bees* Bull. Dep. Agr. W. Aust. 3109

(1964) Beekeeping in Western Australia *Bee World* 45(l): 19-31

(1964) The hive environment in hot climates *J. apic. Res.* 3(2): 117-122

(1964) Responsibilities of Beekeepers under the Beekeepers Act, 1963 *Apiculture* 1: 4-6

(1964) Honey house sanitation *Apic.* 1: 23-25

(1964) Judging honey *Apic.* 1: 38-39

(1964) Pollen substitute *Apic.* 1: 40-41

(1964) Heat reflection by paint *Apic.* 1: 55-56

(1965) *Report on the visit of the Senior Apiculturist to Queensland, New South Wales, South Australia and Victoria* Perth: W.A. Dep. Agr. 17pp.

(1965) *Training in Beekeeping* Perth: W.A. Dep. Agr. Bull. 329ff

(1965) Some considerations in selecting bees for breeding *Aust. Beekeeper* 66(7): 186-189

(1965) W.A. apiculturist looks at beekeeping in the Eastern States *Aust. Beekeeper* 66: 264-268, 283-286

(1965) *Economics of beekeeping* Perth: Dep. Agr. Bull. 3354

(1965) The sucrose content of Western Australian honey *J. apic. Res.* 4(3): 177-184

(1965) Hives in hot climates *Apic.* 1: 86-88

(1966) The cost of protein in pollen substitutes *Aust. Beekeeper* 67(8): 185-186

(1966) The cost of protein in pollen substitutes *Apic.* 1:132-133

(1966) Review on Krawaite research *Aust. Beekeeper* 67(9): 230-231

(1966) The honey super *Gleaning* 94(7): 394-396

(1966) Cappings reducer: a new design *Amer. Bee J.* 106(9): 333-335

(1966) *The Hive* Perth: Dep. Agr. Bull. 3464

(1966) Refractive index and water content of honey *Apic.*1: 164-166

(1967) Honey quality standards *Apic.* 2: 4-5

(1967) Deterioration of the colour of honey *J. Apic. Res.* 6(2): 95-98

(1967) The Honey Super *Apic.* 1: 6-9

(1967) Brood diseases *Apic.* 2: 21-32

(1967) Queen raising *Apic.* 2: 39-43

(1967) Comb honey in sections Apic. 2: 56-59

(1967) Introduction of queens *Apic.* 2: 60-61

(1967) Hot water circulating system *Apic.* 2:69-73

(1968) Economics of honey production *Apic.* 2:82-85

(1968) Honey quality *Apic.* 2: 86-87

(1968) Deterioration of the colour of honey *Apic.* 2: 88-92

(1968) The story of honey *Apic.* 2: 93-96, 92

(1968) *An introduction to beekeeping in Western Australia* Revised edition Perth: Dep. Agr. Bull. 3108 (Reprinted 1971)

(1968) Standard for honey in overseas markets *Apic.* 2: 100-104

(1968) A honey sump strainer *Apic.* 2: 106-109

(1968) "Dyads" in the Proteaceae *Grana Palynologica* 8:86-87

(1969) Wasted honey crops *Apic.* 2: 132-137

(1969) Selecting bees for breeding *Apic.* 2: 146-152

(1969) Dar es salaam beeswax *Apic.* 2: 163-166; 3: 4-5

1969) *Honey plants in Western Australia* Perth: Dep. Agr. Bull. 3618

(1970) The Gantry loader *Apic.* 3: 57-63

(1972) *Vegetation map: Pemberton & Irwin Inlet* Scale 1:250,000 Perth: W. Aust. Dep. Agr.

(1973) *Field for improvement in the techniques of honey production and extraction* 1st Australian Bee Congress, Broadbeach. Bucharest: Apimondia :203-206

(1973) *Vegetation map: Busselton & Augusta* Scale 1:250,000 Perth: W. Aust. Dep. Agr.

(1974) *Vegetation map: Collie* Scale 1:250,000 Perth: W. Aust. Dep. Agr.

(1975) The beekeeper's part in producing honey – Chap. 3 in *Honey: A comprehensive study* London: Heinemann

(1977) *National Park Management Policies* Perth: National Parks Authority pp.31

(1983) *Apicultural Research in Thailand* Perth: W. Aust. Dep. Agr. pp.64

4. AFTER RETIREMENT

One Gunner's War, (1990) Nedlands: FGS pp. 144, photos 18

The Last Cavalry Charge (Supplement to *One Gunner's War*)

Three Cells of Honeycomb, (1994) Nedlands: FGS pp.248, photos 22

Pelinta, (1998) Nedlands: FGS pp.191, photos 30 HB (Building and sailing a trimaran)

Smith, (2001) Nedlands: FGS pp.244, photos 23 HB (Family history)

One Gunner's War 1939-1946, (2002) Nedlands: FGS pp. 169, photos 30 HB

Sixty-Nine Years Together, (2011) Ely: Melrose Books pp .396, photos 16

One Gunner's War, (2012) Ely: Melrose Books pp. 168, photos 26, maps 10

www.ingramcontent.com/pod-product-compliance
Lightning Source LLC
Chambersburg PA
CBHW080322170426
43193CB00017B/2876